„Advances in Polymer Science/Fortschritte der Hochpolymeren-Forschung"

erscheinen zwanglos in einzeln berechneten Heften, die zu Bänden vereinigt werden.

Sie enthalten Fortschrittsberichte monographischen Charakters aus dem Gebiet der Physik und Chemie der Hochpolymeren mit ausführlichen Literaturzusammenstellungen. Sie sollen der Unterrichtung der auf diesen Gebieten Tätigen über solche Themen dienen, die in letzter Zeit besondere Aktualität gewonnen haben, bzw. die in neuerer Zeit eine lebhafte und nach literarischer Zusammenfassung verlangende Entwicklung erfahren haben.

Es ist ohne ausdrückliche Genehmigung des Verlages nicht gestattet, photographische Vervielfältigungen, Mikrofilme, Mikrophoto u. ä. von diesem Heft, von einzelnen Beiträgen oder von Teilen daraus herzustellen.

Anschriften der Herausgeber:

Prof. Dr. H.-J. Cantow, Institut für makromolekulare Chemie der Universität, 78 Freiburg i. Br., Stefan-Meier-Str. 31

Dr. G. Dall'Asta, Istituto di Chimica Industriale del Politecnico, Milano/Italien.

Prof. D. J. D. Ferry, Department of Chemistry, The University of Wisconsin, Madison 6, Wisconsin/USA

Prof. Dr. W. Kern, Institut für Organische Chemie der Universität, 65 Mainz.

Prof. Dr. G. Natta, Istituto di Chimica Industriale del Politecnico, Milano/Italien.

Prof. Dr. C. G. Overberger, Polytechnic Institute of Brooklyn, 333 Jay Street, Brooklyn 1, New York/USA

Prof. D. W. Prins, Laboratorium voor Fysische Chemie, Technische Hogeschool, Delft/Holland

Prof. Dr. G. V. Schulz, Institut für physikalische Chemie der Universität, 65 Mainz

Dr. William P. Slichter, Bell Telephone Laboratories Incorporated, Chemical Physics Research Department, Murray Hill, New Jersey 07971/USA

Prof. Dr. A. J. Staverman, Hugo de Grootstraat 27, Laboratorium voor Anorg. en Phys. Chemie der Rijksuniversiteit Leiden, Leiden/Holland

Prof. Dr. J. K. Stille, University of Iowa, Department of Chemistry, Iowa City/USA

Prof. Dr. H. A. Stuart, Institut für physikalische Chemie der Universität, 65 Mainz

---

4. Band   Inhaltsverzeichnis   4. (Schluß-) Heft

ADVANCES IN POLYMER SCIENCE

FORTSCHRITTE DER HOCHPOLYMEREN-FORSCHUNG

HERAUSGEGEBEN VON

H.-J. CANTOW
FREIBURG I. BR.

G. DALL'ASTA
MILANO

J. D. FERRY
MADISON

W. KERN
MAINZ

G. NATTA
MILANO

C. G. OVERBERGER
NEW YORK

W. PRINS
DELFT

G. V. SCHULZ
MAINZ

W. P. SLICHTER
MURRAY HILL

A. J. STAVERMAN
LEIDEN

J. K. STILLE
IOWA CITY

H. A. STUART
MAINZ

4. BAND

MIT 145 ABBILDUNGEN

SPRINGER-VERLAG
BERLIN HEIDELBERG GMBH

1965—1967

ISBN 978-3-540-03705-7   ISBN 978-3-540-34906-8 (eBook)
DOI 10.1007/978-3-540-34906-8

Library of Congress Catalog Card Number 61-642. Herstellung: Brühlsche Universitätsdruckerei Gießen

Titel Nr. 4918 — 4921

Inhalt des 4. Bandes

Adv. Polymer Sci., Vol. 4, pp. 457—495 (1967)

Thermodynamics of Polymerization with Special Emphasis on Living Polymers

By

M. SZWARC

Department of Chemistry, State University College
of Forestry at Syracuse University
Syracuse, New York 13210

With 14 Figures

Table of Contents

1. General principles

The first basic approach to the thermodynamics of addition polymerization was presented in 1948 by DAINTON and IVIN (1) and developed in their review paper (2) published ten years later. In their exposition, they stressed the significance of the propagation step in addition polymerization, emphasizing its critical role in the whole process. This is the step whereby the macromolecule is gradually formed by the sequence of reactions

$$-M_n - X + M \rightleftarrows -M_{n+1} - X$$

converting the monomer M into polymer molecule $-M_n \cdot X$ possessing the active end group X. For large values of n, i.e. for a high-molecular weight polymer, the reactivity of X is independent of n and then the above equation may be symbolically represented by

$$M_f \rightleftarrows M_s ,$$

30

where M_f and M_s denote, respectively, a free monomer molecule and a monomer segment located somewhere in the midst of a long chain of a polymer molecule $-M_n \cdot X$. The free energy change, ΔF_p, accompanying the addition of one mole of monomer to a high-molecular weight polymer is given, therefore, by the difference $\Delta F_p = \Delta F(M_s) - \Delta F(M_f)$ where $\Delta F(M_f)$ and $\Delta F(M_s)$ are, respectively, the molar free energies of the monomer and of a monomer segment of a high-molecular weight polymer (both referring to the conditions prevailing in the system). ΔF_p is, thus, independent of the nature of X and of the magnitude of n. Its value depends, however, on the nature of the solvent present in the system and, to some extent, on the monomer and polymer concentrations because their presence affects $\Delta F(M_f)$ and $\Delta F(M_s)$ by modifying the environment in which the process takes place. Finally, it should be noted that the free energy of propagation for a specific monomer-polymer system under defined thermodynamic conditions, e.g., pure liquid monomer $\rightarrow$ amorphous solid polymer, is constant regardless of the mechanism of polymerization, provided that the state and conformation of the polymer do not change.

The last corollary deserves some comment. The nature of the growing end and the mechanism of polymerization affect the rates of propagation and depropagation. However, thermodynamic functions are always independent of the reaction path and are determined uniquely by the initial and final states of the system. Hence, any conceivable process, even if impractical, may serve to calculate ΔF_p. Let us consider, e.g., a high-molecular weight polymer with the terminal X and Y groups linked by a long chain of n monomer segments

$$X.M.M.\ldots\ldots\underbrace{M.M}\ldots\ldots M.M.Y.$$
$$\underbrace{\qquad\qquad}_{\text{long chain}} \quad \underbrace{\qquad\qquad}_{\text{long chain}}$$

The conversion of an n-mer into an $(n + 1)$-mer may be imagined to occur in two steps: (1) fission of an $M-M$ bond located in the middle of the chain; (2) insertion of a monomer molecule into the broken linkage coupled with the formation of two new $M-M$ bonds. Because the end groups are far away from the reaction center, their nature cannot influence the free energy of the process which is therefore independent of the magnitude of n and of the nature of X and Y. This remains true even if X and Y are some unreactive end groups of a "dead" polymer.

The conventional thermodynamic equation

$$\Delta F_p = \Delta H_p - T\Delta S_p$$

relates the free energy of propagation to the heat and entropy change of this process. Depending on the signs of ΔH_p and ΔS_p, four distinct situations may occur:

(1) For negative ΔH_p and ΔS_p, (most commonly the case encountered in addition polymerization) ΔF_p becomes positive above a certain critical temperature, $T_c = \Delta H_p / \Delta S_p$, known as the "ceiling temperature" of the system. The value of T_c depends on the concentrations of the monomer and of the polymer as well as on the nature of the solvent, if the latter is present in the system. Of course, the high polymer cannot be formed above the ceiling temperature. For any monomer-polymer system, the process which converts pure liquid monomer into a crystalline polymer has the maximum ceiling temperature.

(2) For positive ΔH_p and ΔS_p, the system can polymerize only above the critical floor temperature, $T_f = \Delta H_p / \Delta S_p$. The phenomenon of floor temperature is exhibited by some cyclic monomers. On polymerization they lose part of the binding energy of the ring associated with the closeness of some atoms or groups which become separated upon forming the linear structure. Polymerization of such monomers may increase the entropy of the system, if the loss of monomer translational and rotational entropy is more than compensated for by the entropy of internal rotation around those bonds which were rigid in the original cyclic molecule but allow free rotation in the newly formed chain. The polymerization of octameric sulfur molecules, S_8, into polymer chains of plastic sulfur exemplifies such a process (for further details see p. 486).

It should be remarked that dilution of the monomer decreases the entropy of polymerization and, hence, even if its value were positive for pure monomer it eventually becomes negative at a sufficiently high dilution. Therefore, for any system showing the phenomenon of floor temperature, polymerization becomes impossible at any temperature below a certain critical monomer concentration.

(3) When ΔH_p is positive and ΔS_p negative, no polymerization is possible under any conditions.

(4) When ΔH_p is negative and ΔS_p positive, polymerization may occur at all temperatures.

In some systems, the same polymer may be produced from different monomers, e.g., plastic sulfur can be formed by polymerizing S_6 as well as S_8 molecules. In such systems, a polymer may be stable with respect to one monomer but labile with respect to another one. For example, linear siloxane $(-Me_2SiO-)_n$, which could be formed from the cyclic tetramer $(-Me_2SiO-)_4$, decomposes into the cyclic trimer $(-Me_2SiO-)_3$ and the reaction $3\,(-Me_2SiO-)_4 \rightarrow 4\,(-Me_2SiO-)_3$ proceeds, therefore, through a polymer intermediate.

Similar phenomena may be observed in some copolymerizations. For example, tetrafluoroethylene, C_2F_4, and trifluoronitrosomethane, CF_3NO, copolymerize spontaneously into a $1:1$ copolymer, $-[-CF_2.CF_2.O.N(CF_3).-]_n-$. Its degradation produces, however, a pair of inert monomers, i.e.

$$\text{---}[O.CF_2.CF_2.N(CF_3)]_n \text{---} \rightarrow n(CF_2O) + n(CF_2:NCF_3)$$

which are thermodynamically more stable than the copolymer.

2. Experimental determination of ceiling temperature

The ceiling temperature of a polymerizable system may be obtained from studies of its polymerization kinetics at a series of rising temperatures. For many polymerizations, the sum of the activation energies

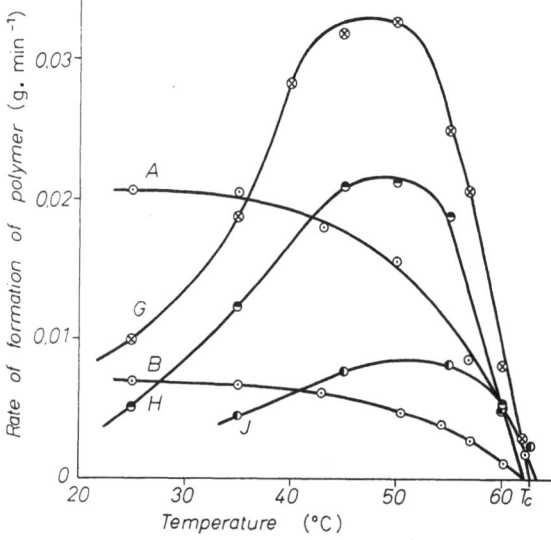

Fig. 1. Rate of formation of polybutene sulfone from monomer mixtures containing 9,1 mole-% 1-butene. (*A*) and (*B*) photochemical initiation at two different intensities; (*G*) and (*H*) initiation by silver nitrate at two different concentrations; (*J*) initiation by benzoyl peroxide. Reproduced, with permission, from Dainton and Ivin; Disc. Faraday Soc. **14**, 199 (1953).

for the initiation and propagation steps greatly exceeds the activation energy for the termination step and, therefore, the initial rate increases with temperature. However, as the ceiling temperature is approached, the depropagation step, which at the lower temperatures is too slow to contribute significantly to the rate, becomes important and the rate decreases steeply as shown in Fig. 1. Extrapolation to zero rate gives, therefore, the required ceiling temperature. The accuracy of the method dependens on the steepness of the descending line, and usually it is good to about 2—3° C.

The kinetic technique was extensively used by Dainton and his associates (*3, 4*) particularly in their studies of the copolymerization of olefins with sulfur dioxide to 1 : 1 polysulfones. To check their results, they compared the heats of polymerization calculated from the ceiling temperatures determined at various partial pressures of olefin and sulfur-dioxide with those directly obtained by calorimetry. The agreement between these two sets of data is shown in Table 1. Further verification (5) was provided by calorimetric determinations of the respective ΔS and comparing it with $\Delta S = \Delta H/T_c$.

Table 1. *Heats of 1:1 copolymerization of olefins and SO_2 to polysulfones*

Olefin	$-\Delta H$ (from T_c) Kcal./Mole	$-\Delta H$ (from Calorimetry) Kcal./Mole
1-Butene	20.7 ± 1.4	21.2 ± 0.1
cis-2-Butene	20.8 ± 0.7	20.1 ± 0.1
trans-2-Butene	19.3 ± 0.5	18.7 ± 0.1
1-Hexadecene	19.2 ± 1.2	19.9 ± 0.1

Alternatively, one might investigate the reaction between polymer free radicals and monomer vapor. Chains of polymer molecules, kept in a monomer atmosphere, are ruptured by a suitable photochemical technique, thus producing the respective polymer radicals. These undergo polymerization and consume monomer, if its vapor pressure exceeds the

Table 2. *Heats of polymerization determined from ceiling temperatures and from calorimetry*

Monomer	$-\Delta H$ (from T_c) Kcal./Mole		$-\Delta H$ (from calorimentry) Kcal./Mole	
Methyl mathacrylate	13.4 ± 0.5	(a, b)	13.9 ± 0.3	(c)
Ethyl methacrylate	14.4 ± 0.6	(d)	14.1	(e)
Methacrylonitrile	15.3 ± 1.0	(f)	—	
α-Methylstyrene	8.15	(g)	8.42	(h)
$\in$-Caprolactam	3.6	(i)	3.25	(j)

(a) S. BYWATER: Trans. Faraday Soc. **51**, 1267 (1955).
(b) K. J. IVIN: Trans. Faraday Soc. **51**, 1273 (1955).
(c) S. EKEGREN, O. OHRN, K. GRANATH, and P. O. KINELL: Acta Chem. Scand. **4**, 126 (1950).
(d) R. E. COOK, and K. J. IVIN: Trans. Faraday Soc. **53**, 1132 (1957).
(e) K. IWAI: J. Chem. Soc. Japan, Ind. Chem. Sct. **49**, 185 (1946).
(f) S. BYWATER: Can. J. Chem. **35**, 552 (1957).
(g) S. BYWATER, and D. J. WORSFOLD: J. Polymer Sci. **26**, 299 (1957).
(h) D. E. ROBERTS, and R. S. JESSUP: J. Res. Natl. Bur. Std. **46**, 11 (1951).
(i) A. B. MEGGY: J. Chem. Soc. 796 (1953).
(j) S. M. SKURATOV, A. A. STREPIKHEEV, and E. N. KANARSKAYA: Kolloidn. Zh. **14**, 185 (1952).

respective equilibrium value, p_e whereas depolymerization takes place and free monomer is liberated in the reverse case. Consequently, the pressure in the system decreases when $p > p_e$ but increases when $p < p_e$ and, thus, at each temperature the respective equilibrium pressure may be found by interpolation. This method was used by IVIN (6) in his studies of the polymethylmethacrylate-gaseous methyl methacrylate system and it was subsequently extended to other systems listed in Table 2. A similar approach, adapted to equilibria established in solutions, was developed by BYWATER (7).

3. Thermodynamic studies of polymerizations involving living polymers

The equilibrium techniques of IVIN and of BYWATER, discussed above' can be applied elegantly to systems involving living polymers. The ability of these species to grow implies, in accordance with the principle of microscopic reversibility, their ability to degrade. Hence, in a living polymer-monomer system, the following equilibria must eventually be established:

$$P_{n_0}^* + M \rightleftarrows P_{n_0+1}^*, \ldots K_{n_0}$$

$$P_{n_0+1}^* + M \rightleftarrows P_{n_0+2}^*, \ldots K_{n_0+1}$$

$$\cdots \cdots \cdots \cdots \cdots \cdots \cdots$$

$$P_{n_0+i-1}^* + M \rightleftarrows P_{n_0+i}^*, \ldots K_{n_0+i-1}$$

$$\cdots \cdots \cdots \cdots \cdots \cdots \cdots$$

In these equations, $P_{n_0+i}^*$ denotes a living n_0+i-mer, M the monomer, and $P_{n_0}^*$ the lowest living polymer which may grow but not degrade. For example, in the cumylpotassium-α-methylstyrene system, $n_0 = 1$ and $P_{n_0}^*$ is cumylpotassium. It is possible to convert α-methylstyrene into a tail-to-tail dimer

$$K^+, {}^-C(Me)(Ph) \cdot CH_2 \cdot CH_2 \cdot C^-(Me)(Ph), K^+ = ({}^-\alpha\alpha^-)$$

and therefore in the system ${}^-\alpha\alpha^- + \alpha$ (α denoting α-methylstyrene) the dimer acts as $P_{n_0}^*$.

Whenever $K_{n_0} = K_{n_0+1} = \cdots = K_{n_0+i} = \cdots = K_\infty$, the equilibrium concentration of the living n_0+i-mer, $P_{n_0+i}^*$, may be expressed in terms of the equilibrium concentrations of the monomer, M_e, and of the living n_0-mer, $P_{n_0}^*$, namely,

$$P_{n_0+i}^* = P_{n_0}^* (K_\infty M_e)^i .$$

TOBOLSKY (8) has shown that in such a case M_0, P_{total}^* and M_e are related by the following equation:

$$\{M_0 - M_e\}/P_{total}^* = K_\infty M_e/\{1 - K_\infty M_e\}$$

where M_0 is the initial concentration of the monomer which was added to a solution of the living n_0-mers present initially at concentration P_{total}^*.

Evidently, $P_{total}^* = \sum_0^\infty P_{n_0+i}^*$ and $(M_0 - M_e)$ is the amount of monomer polymerized per unit volume of reacting solution. The total volume of the solution was assumed implicitly to be constant and not affected by the polymerization — a reasonable assumption for dilute solutions. The problem of volume contraction may, however, be circumvented (9) by expressing all the concentrations in moles per unit mass of solution.

TOBOLSKY's approach requires some modification if K_j varies with j. In a more general treatment, described recently by SZWARC (10), the equalities $K_j = K$ are assumed to be valid only for j's exceeding some value $s + 1$ ($s > 1$), whereas for $j \leq s$ the respective K_j's may differ from K_∞ and from each other. In such a system, K_∞ may be found from the equation

$$(M_0 - R_s - M_e)/(P^*_{total} - Q^*_s) = K_\infty M_e/(1 - K_\infty M_e)$$

where the symbols R_s and Q^*_s are defined by the ratios Q^*_s/P^*_{total} and R_s/M_0. The former represents the mole fraction of all the living polymers having degress of polymerization less than $n_0 + s$, whereas the latter denotes the fraction of the initially introduced monomer which becomes incorporated into the above polymers of DP $\leq n_0 + s$ (DP = degree of polymerization). All the remaining symbols retain their previous meaning.

For a high average degree of polymerization, the ratios Q^*_s/P^*_{total} and $(P^*_{total} - Q^*_s)/(M_0 - R_s - M_e)$ tend towards zero and, thus, $K_\infty = \lim. (M_e^{-1})$. The determination of K_∞ is reduced, therefore, to an analytical problem of finding the ultimate concentration of the monomer which coexists in equilibrium with its high-molecular weight living polymer. In some systems, notably α-methylstyrene, the analysis presents no difficulties because the monomer equilibrium concentration is relatively high, e.g., in a tetrahydrofuran solution of living poly-α-methylstyrene, $M_e = 0.8$ mole/liter at $0°$ C. This system was thoroughly investigated by MCCORMICK (11), by WORSFOLD and BYWATER (12), and by VRANCKEN, SMID, and SZWARC (13). Their combined results are presented in a plot of $\log K_\infty$ against $1/T$ as shown in Fig. 2. The agreement between all three sets of data is excellent and the best line through all the experimental points gives $\Delta H = -7.5$ kcal./mole and $\Delta S = -26.5$ e.u.

MCCORMICK (11) and WORSFOLD and BYWATER (12) initiated the polymerization of α-methylstyrene by using sodium naphthalene, brought the system to equilibrium at the desired temperature, and then "killed" the living polymers. They showed that at $0°$ C the concentration of the monomer attains its equilibrium value in less than 16 hours, and they observed no further polymerization over periods lasting for as long as 124 hours. Furthermore, they showed that the monomer equilibrium concentration was independent of the concentration of living ends, i.e. of the amount of catalyst applied.

Various problems may be encountered in thermodynamic investigations of living polymers. An instructive example is provided by the studies of the styrene-living polystyrene system as reported recently by

Bywater and Worsfold (*14*). At 0° C, the equilibrium concentration of styrene was expected to be about 10^{-7} mole/liter which is too low to be determined by conventional analytical techniques. The system was, therefore, investigated in the temperature range of 100–150° C, where the equilibrium concentrations were expected to rise to 10^{-4}–10^{-3} mole/liter. For these ultraviolet spectrophotometric techniques are applicable. This temperature range is well above that normally considered

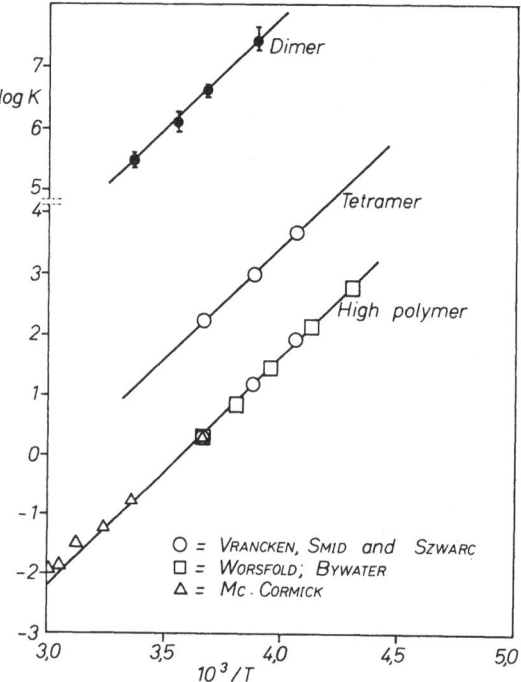

Fig. 2. Equilibrium constant for polymerization of α-methylstyrene to living oligomers and to a high molecular weight polymer. (O) Vrancken, Smid, and Szwarc; (□) Worsfold and Bywater; (△) McCormick. Reproduced, with permission, from Vrancken, Smid, and Szwarc; Trans. Faraday Soc. **58**, 2036 (1962).

suitable for anionic polymerizations. In fact, in tetrahydrofuran under these conditions one observes a rapid "isomerization" of living polystyrene as manifested by the appearance of a new absorption band at $\lambda_{max} = 560$ mμ and by a decay of the optical density at $\lambda_{max} = 340$ mμ, the absorption band characteristic of living polystyrene. To avoid these difficulties, the equilibrium was studied in hydrocarbon solvents, i.e. in benzene and in cyclohexane, using butyllithium to initiate the polymerization. At these elevated temperatures, the half-life of the reaction was found to be of the order of a few seconds in cyclohexane and even less in benzene and, hence, in less than a minute the system approached, within 1%, its equilibrium state. This was important because some slow side reactions still take place even in these solvents, e.g., the characteristic 334 mμ absorption band of lithium-containing living polystyrene, ⸺$CH_2.CH(Ph)^-$, Li^+, decreased during the experiments and a small absorption band appeared near 450 mμ. Fortunately, the half-life of this side reaction was nearly a hundred times longer than that of the polymerization and, thus, this disturbance was of no significance unless the solution was heated for an excessively long time. The equilibrium was quenched by adding a small amount of water at the temperature of the experiment. Thereafter, the mixture was cooled rapidly to minimize any thermal polymerization of the residual monomer.

The results, collected in Table 3 are shown graphically in Fig. 3 as a plot of $\log K_\infty$ against $1/T$. Benzene and cyclohexane are good and poor

Table 3. *Equilibrium concentration M_e of styrene in contact with living polystyrene*[1]

$T°C$	M_e in Benzene Mole/Liter $\times 10^5$	M_e in Hexane Mole/Liter $\times 10^5$
100	—	3.97
110	12.0	7.79
120	20.7	13.6
130	34.2	23.8
140	59.0	41.6
150	90.8	65.0

[1] From S. Bywater, and D. J. Worsfold, J. Polymer Sci. **58**, 571 (1962).

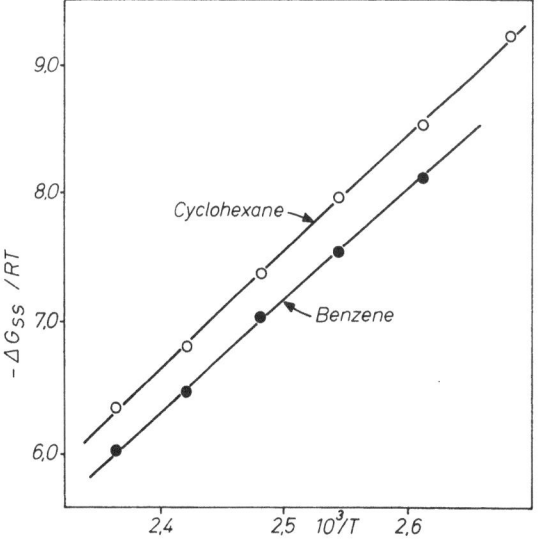

Fig. 3. Equilibrium constant of living polystyrene propagation in cyclohexane and benzene; $\ln K = -\Delta G_{ss}/RT$ plotted versus $1/T$. Reproduced, with permission, from Bywater and Worsfold: J. Polymer Sci **58**, 571 (1962).

solvents, respectively, for polystyrene and their effects upon the equilibrium polymerization constants are apparent from the displacement of the lines as shown in Fig. 3. The difference in the respective K_∞'s arises mainly from the polymer-solvent interactions which affect the free energy of the polymer segments. A more extensive discussion of this subject will be found in section 11.

The living polymer technique has been applied recently in studies of equilibria tetrahydrofuran $\rightleftarrows$ polytetrahydrofuran. This polymerization proceeds by a cationic mechanism (15) the propagation being described by the equation

$$-\!\!-\!\!-O\cdot(CH_2)_4\!-\!\underset{X^-}{\overset{+}{O}}\!\!\underset{CH_2\!-\!CH_2}{\overset{CH_2\!-\!CH_2}{\diagdown}} + O\!\!\underset{CH_2\!-\!CH_2}{\overset{CH_2\!-\!CH_2}{\diagdown}} \rightleftarrows$$

$$-\!\!-\!\!-O\cdot(CH_2)_4\!-\!O\!-\!(CH_2)_4\!-\!\underset{X^-}{\overset{+}{O}}\!\!\underset{CH_2\!-\!CH_2}{\overset{CH_2\!-\!CH_2}{\diagdown}}$$

Various agents may initiate such a polymerization, e.g., Ph_3C^+, $SbCl_6^-$, whose action was investigated by Bawn and his coworkers (16). They also were first to recognise the formation of living polymers in tetrahydrofuran

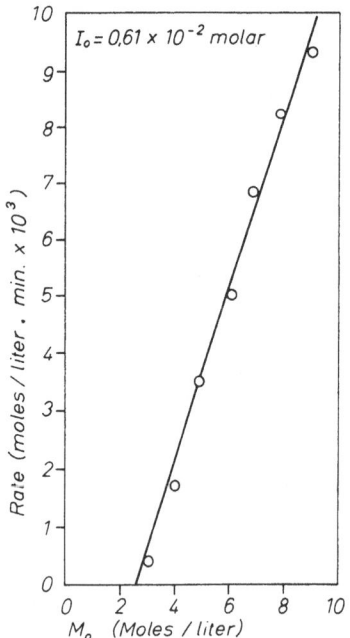

Fig. 4. Rate of cationic polymerization of tetrahydrofuran to polyether at 0° C as function of monomer concentration. Reproduced, with permission, from Vofsi and Tobolsky: J. Polymer Sci. 3 A, 3261 (1965).

system. Vofsi and Tobolsky (17) investigated the kinetics of tetrahydrofuran polymerization initiated by Et_3O^+, BF_4^- and determined the monomer equilibrium concentration to be 2.6 M at 0° C. They showed that the initial rate of polymerization decreases linearly with the initial monomer concentration, see Fig. 4, the extrapolation to zero initial rate giving the equilibrium monomer concentration. Conceptually this approach is similar to that developed by Ivin (6).

A most thorough investigation of the tetrahydrofuran system was recently reported by Dreyfuss and Dreyfuss (18). They initiated the polymerization by the decomposition of benzenediazonium hexafluorophosphate (PhN_2^+, PF_6^-), which provides a Ph^+ ion and an extremely stable PF_6^- counterion. It seems that the stability of the counterion is the reason for the simplicity of the system which is admirably suitable for the kinetic and thermodynamic studies. The existence of living polymers was demonstrated in a variety of ways, e.g., by the increase of the molecular weight of the polymer on the addition

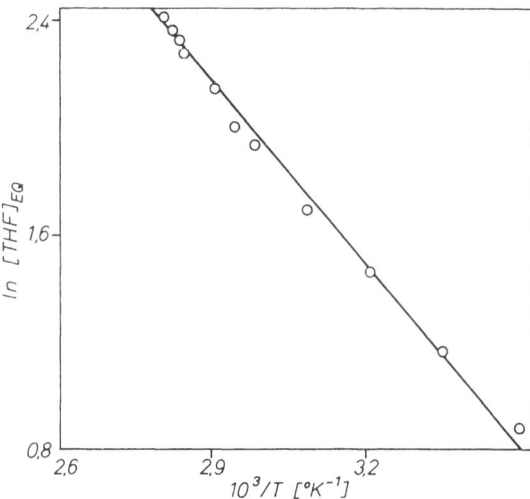

Fig. 5. Equilibrium constant of tetrahydrofuran polymerization to polyether as function of temperature (medium pure tetrahydrofuran). Reproduced, with permission, from DREYFUSS and DREYFUSS: J. Polymer Sci. 4A, 2179 (1966).

of further monomer, by the formation of block polymers, etc. The equilibrium concentration of the monomer was determined at various temperatures and the plot of $\log K_e$ against $1/T$ is shown in Fig. 5.

4. Use of radioactive tracers in the determination of the equilibrium concentrations of monomers

For many systems the equilibrium concentration of monomer is expected to be extremely low and one may attempt to use the radioactive tracer technique to determine its value. The method requires some refinements because even a small isotope effect might affect considerably the final result. The following calculation illustrates this point.

Let x_0 and x_e denote the respective initial and final (equilibrium) concentrations of the monomer and a_0 and a_f their initial and final specific activities, expressed as mole fraction of the radioactive tracer. Consider an isotope effect which makes the rate of addition of the tracer molecule f-times greater than that of the non-radioactive monomer. Hence, the increase, da, in the specific activity of the residual monomer, arising from the polymerization of $-dx$ moles of monomer $(dx < 0)$, is given by

$$a + da = a(x + fdx)/(x + dx)$$

where x and a denote the respective concentration and specific activity of the monomer at some stage of polymerization. Rearrangement of this equation leads to

$$da/a = (f - 1)\, dx/x$$

which on integration gives

$$\log (a_f/a_0) = (f - 1) \log (x_e/x_0)\,.$$

Denoting the fraction of the total activity recovered in the residual monomer by γ, so that $\gamma = x_e a_f / x_0 a_0$, one finds

$$\log \gamma = f \cdot \log x_e - f \cdot \log x_0\,.$$

Hence, a plot of $\log \gamma$ against $\log x_0$ should result in a straight line having a slope, $-f$, and an intercept, $f \cdot \log x_e$.

In any real system, f is expected to deviate from unity by not more than 10%. Two numerical examples will show how much the results are affected by such a deviation. Consider, e.g., a polymerization which occurs upon mixing a 1 M solution of monomer with living polymers so that the resulting reaction reduces eventually the monomer concentration to its equilibrium value of 10^{-7} mole/liter. Using the derived formulae, we find

$$a_f/a_0 = 5.0 \quad \text{for} \quad f = 0.90$$
$$a_f/a_0 = 1.2 \quad \text{for} \quad f = 0.99\,.$$

Hence, even for an isotope effect as low as 1%, the specific activity of the residual monomer would increase by about 20%. The enrichment of monomer with radioactive tracer affects the apparent x_e values. For example, had the correction not been introduced in the calculation, erroneous values of $x_e = 5.0 \cdot 10^{-7}$ mole/liter or $1.2 \cdot 10^{-7}$ mole/liter (for $f = 0.90$ and 0.99, respectively) would be obtained instead of the correct value of $x_e = 1.0 \cdot 10^{-7}$ mole/liter.

In deriving these equations, exchange between monomer and living polymer was ignored. This is permissible when the rate of depropagation is much slower than the rate propagation. However, as the system approaches equilibrium, the depropagation step becomes comparable to the propagation step and the validity of the approximation may then be questioned. Fortunately, the final results are only slightly distorted by this factor, because the specific activity of the monomer varies slowly with its concentration. In the example given, for $f = 0.90$, the specific activity of the monomer increased fourfold as its concentration decreased from 1 mole/liter to 10^{-6} mole/liter. At this stage, propagation is still ten times faster than depropagation. Further decrease in the monomer concentration to the limiting value of 10^{-7} mole/liter increases its specific activity only by an additional 25%. Hence, at this last reaction stage, the specific activity of the growing ends (and not of the polymer as a whole) is similar to the expected final, specific activity of the monomer so that the exchange cannot be of any great importance, if the mixture is not left for too long before "killing" the polymers.

To complete this discussion, let us consider the plot of $\log \gamma$ against $\log x_0$ as shown in Fig. 6. Evidently, our treatment applies only when x_0 is much greater than x_e, but for x_0 comparable to x_e, the theoretical straight line becomes curved as illustrated in Fig. 6. For ratios x_0/x_e of

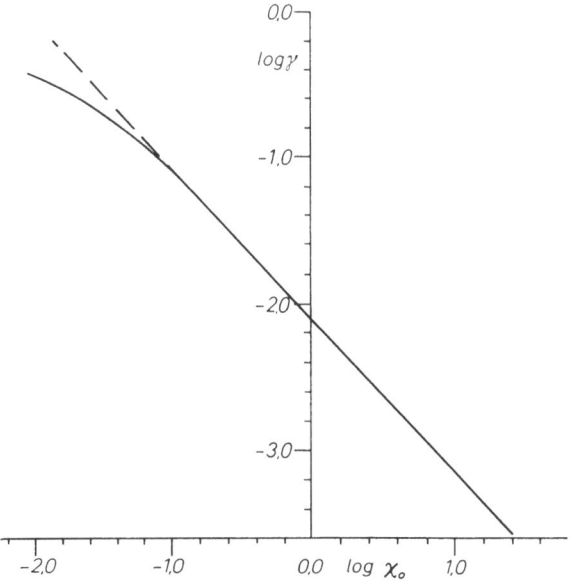

Fig. 6. Dependence of the recovered activity fraction of a radioactive monomer, remaining in equilibrium with its living polymer, on its initial concentration. $K_e = 100$ moles/liter, $P^* = 0.01$ moles/liter, $f = 1.05$.

1000 or more, the deviations from the theoretical line are expected to be small and the tracer method may be used whenever $x_0/$[growing ends] is sufficiently large, e.g. greater than 100.

It should be stressed that this technique may be utilized for the determination of small isotope effects in polymerization. In fact, it is more reliable for this purpose than for determining equilibrium concentration of the monomer.

5. Equilibria between monomer and living oligomers

Although for high-molecular weight polymers, the equilibrium propagation constant is independent of the degree of polymerization, this need not be the case for low-molecular weight oligomers, i.e. dimers, trimers or tetramers, or for macromolecules which involve head-to-head or tail-to-tail linkages. In fact, recent studies of oligomers of α-methylstyrene (13) revealed great deviations of such equilibrium constants from the K_∞ values.

Prolonged contact of α-methylstyrene with potassium or sodium-potassium alloy in tetrahydrofuran yields a living, tail-to-tail dimer, i.e.

$$K^+, {}^-C(CH_3)(Ph).CH_2.CH_2.C(CH_3)(Ph)^-, K^+ = K^+, {}^-\alpha\alpha^-, K^+.$$

This structure was conclusively established by carboxylation of the dimer which produced a mixture of stereoisomers of 2-5-dimethyl-2,5-diphenyl-adipic acids (19).

It will be shown later (see section 6) that the addition of α-methylstyrene to both ends of the dimer proceeds much faster than further growth of the resulting tetramer. Consequently, a living tetramer, denoted by T_2, may be prepared by the addition of the required amount of α-methylstyrene to the living α-methylstyrene dimer ($K^+, {}^-\alpha\alpha^-, K^+$). The structure of T_2 follows from the method of preparation, namely,

$$T_2 = K^+, {}^-C(CH_3)(Ph).CH_2.C(CH_3)(Ph).CH_2.CH_2.C(CH_3)(Ph).$$
$$.CH_2.C(CH_3)(Ph)^-, K^+$$

or symbolically, $T_2 = {}^-\text{HHHH}^-$.

On the other hand, relatively brief contact of a diluted tetrahydro-furan solution of α-methylstyrene with metallic sodium gives an isomeric living tetramer (13) denoted by T_1. Studies of its formation (20) indicated a tail-to-tail, head-to-head, tail-to-tail structure for this species, i.e.

$$T_1 = Na^+, {}^-C(CH_3)(Ph).CH_2.CH_2.C(CH_3)(Ph).C(CH_3)(Ph).$$
$$.CH_2.CH_2.C(CH_3)(Ph)^-, Na^+$$

or symbolically, $T_1 = {}^-\text{HHHH}^-$. All these oligomers show the characteristic absorption spectrum of benzyl carbanions, i.e. $\lambda_{max} = 340$ mμ and $\varepsilon = 1 \cdot 10^4$ (per end), leaving no doubt that these species contain terminal $-CH_2.C(CH_3)(Ph)^-$ units.

Formation of tetramers by the action of metallic sodium on ether solutions of α-methylstyrene had been described previously by BERGMANN et al. (21) who proposed the following structure for this compound

$$Na^+, {}^-C(CH_3)(Ph).CH.C(CH_3)(Ph) - CH_3$$
$$|$$
$$Na^+, {}^-C(CH_3)(Ph).CH.C(CH_3)(Ph) - CH_3.$$

This work needs further verification. BERGMANN's conclusion is based on the assumption that through some rearrangement of unknown nature, a dimer

$$C(CH_3)(Ph):CH.C(CH_3)(Ph).CH_3$$

is initially formed, and that it in turn reacts further with sodium giving the proposed product.

The equilibrium constants for the addition of monomer to low-molecular weight living oligomers have been determined by a method developed by VRANCKEN, SMID, and SZWARC (13). In a solution containing a constant, although very low, concentration of living oligomers which may grow but not decompose, the equilibrium concentration, M_e, of an added monomer may be determined as a function of its initial concentration, M_0. A typical result is shown in Fig. 7 where M_e is plotted as a function of M_0 for a constant concentration of living tetramer, T_1. In fact, such a plot proves the stability of the tetramer, T_1, because the

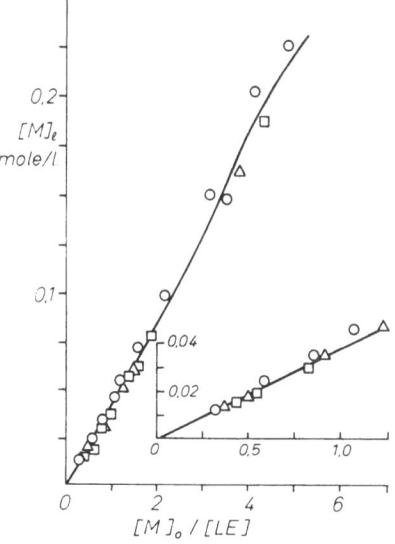

Fig. 7. Equilibrium concentration $[M]_e$ of α-methylstyrene in contact with its living tetramer as a function of its initial concentration M_0. Solvent, THF; temperature, 0° C; concentration units, moles/liter. Results of three series of experiments, each corresponding to about 0,05 M concentration of living tetramer. Reproduced, with permission, from VRANCKEN, SMID, and SZWARC; Trans. Faraday Soc. **58**, 2036 (1962).

resulting curve passes through the origin indicating that T_1 does not decompose into monomer and a trimer. The decomposition of a tetramer would be expected had one of the terminal monomeric units been linked in a head-to-tail fashion. Since dissociation is not observed, this provides an additional argument favoring the proposed tail-to-tail, head-to-head, tail-to-tail structure of T_1. The stability of T_1 may be contrasted with the behavior of tetramer T_2 which is stable only in the presence of an equilibrium concentration of monomer amounting, e.g., at 25° C, to 0.02 mole/liter of α-methylstyrene for the concentration of 0.04 mole/liter of living ends.

Having established the equilibrium concentration of the monomer, M_e, as a function of its initial concentration, M_0, one may proceed to calculate the respective equilibrium constants. VRANCKEN et al. (13) accomplished this as follows: they assumed that equilibrium had been established between monomer and *all* of the living j-mers, $j \gg n_0$, that each end of the two-living ended polymer reacts independently of the fate of the other end, and that *all* of the equilibrium constants have identical values, i.e. $K_1 = K_2 = \cdots = K_\infty$. The apparent constant, K_1', was calculated from Tobolsky's relation

$$(M_0 - M_e)/P^*_{\text{total}} = K_1' M_e (1 - K_1' M_e)^{-1} .$$

In this equation, P^*_{total} denotes the total concentration of living ends which is equal to the initial concentration of the growing ends of the living n_0-mer. Keeping $P^*_{\text{total}} = (P^*_{n_0})_0$ constant, they gradually increased M_0 in a series of experiments. Every pair of values, M_0 and its respective M_e, enabled them to calculate the apparent constant K_1', and hence

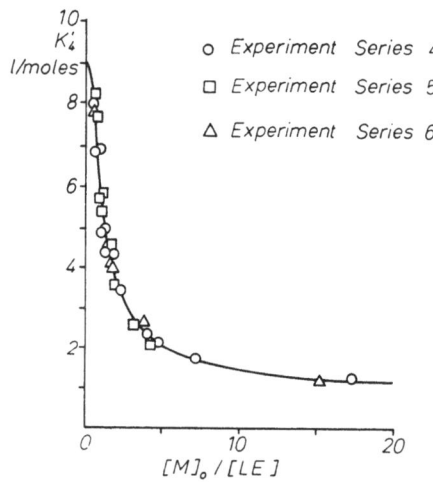

Fig. 8. Apparent equilibrium constant, K_4 for the system living α-methylstyrene-living tetramer in THF at 0° C. (○) Experiment series 4; (□) experiment series 5; (△) experiment series 6. Reproduced, with permission, from Vrancken, Smid, and Szwaro; Trans. Faraday Soc. **58**, 2036 (1962).

K_1' may be determined as a function of M_0. Had the assumption, $K_1 = K_2 = \cdots = K_\infty$ been valid, then each pair of values, M_0 and M_e, would have given the same $K_1' = K_1$, i.e. a plot of K_1' against M_0 should then produce a straight line parallel to the M_0 axis. The authors found (13), however, that such a plot was curved, as shown in Fig. 8, invalidating the assumption requiring *all* K_j to be identical. Nevertheless, the first equilibrium constant, K_1,

$$P^*_{n_0} + M \rightleftarrows P^*_{n_0 + 1}, \ldots K_1$$

may be determined from such a plot by extrapolating the experimental curve to $M_0 = 0$. In this way, Vrancken et al. determined the equilibrium constant for the processes,

$$^-\alpha\alpha^- + \alpha \rightleftarrows \text{Trimer}, \ldots K_D$$

and

$$T_1 + \alpha \rightleftharpoons \text{Pentamer}, \ldots K_T \, ,$$

and the pertinent data are collected in Table 4.

It should be noticed that the function $K_1' = f(M_0)$ asymptotically approaches K_∞ for a high ratio M_0/P_{total}^* and, hence, the latter constant may also be determined from the plots shown in Fig. 8. The same value of K_∞ should be obtained from the data derived from studies either of the dimer or of the tetramer. The results listed in the fourth and fifth columns of Table 4 confirm indeed this conclusion. Moreover, as shown

Table 4. *The equilibrium constant of growth process for living α-methylstyrene oligomers*

$T°C$	K_D, Liter/Mole	K_{T_1} Liter/Mole	K_∞(extrap.K_D'); K_∞ (extrap.K_{T_1}'); K_∞ (lit.)		
				Liter/Mole	
25.4	240 $\pm$ 30	—	0.8 (?)	—	0.41
10.0	450 $\pm$ 50	—	0.85	—	0.9
0.	750 $\pm$ 100	9.4	1.24	1.33	1.35
—15.0	1750 $\pm$ 300	20.	3.3	3.3	3.3
—26.5	—	40.	—	6.8	6.8

K_D refers to $^-\alpha\alpha^- + \alpha \rightleftharpoons {}^-\alpha\alpha\alpha^-$.
K_{T_1} refers to $^-T_1^- + \alpha \rightleftharpoons {}^-T_1\alpha^-$.
K_∞ refers to high-molecular weight poly-α-methylstyrene.
K_∞ (lit.) data from combined results of H. W. McCormick (11) and D. J. Worsfold and S. Bywater (12).

in the last column of this table, the K_∞ values thus derived agree well with those reported by McCormick (11) and by Worsfold and Bywater (12).

The procedure applied to determine K_1 may be utilized to calculate K_2

$$P_{n_0+1}^* + M \rightleftharpoons P_{n_0+2}^* \ldots K_2 \, .$$

By assuming that $K_2 = K_3 = \ldots = K_\infty$ and accepting the K_1 value previously derived, K_2' may be calculated (18), namely,

$$K_2' = M_e^{-1} - (1/2) K_1 \left\{ \sqrt{1 + 4 P_{\text{total}}^* / K_1 M_e (M_0 - M_e)} - 1 \right\} .$$

Extrapolation of K_2' to $M_0 = 0$ should then give the correct value for K_2, while K_∞ would again be given by the asymptotic value of K_2' for a large M_0/P_{total}^*. An extension of such calculations should give values for K_3, K_4, etc. However, the experimental errors accumulate rapidly in such calculations, rendering the method useless for the determination of K_3 and unreliable even for K_2, if the individual errors of experimental M_e values exceed 2%.

The preceding discussion brings us to the general problem of the reliability of such determinations of equilibrium constants. The reproducibility of the data may be seen by examining plots of M_e as a function of M_0 for a constant P^*_{total} as shown, e.g., in Fig. 7. By this test, the results seem to be satisfactory. However, much more severe demands on the reproducibility and the reliability of the data are imposed by plotting K'_1 as a function of M_0 and Fig. 8 illustrates this point. While K'_1 is a slowly varying function of M_0 for large values of M_0/P^*_{total}, the curve rises steeply as M_0/P^*_{total} approaches zero. The steepness of such curves increases as the ratio K_1/K_∞ becomes larger and, hence, the reliability of the data must be decidedly greater in studies of the living dimer $(K_D/K_\infty \approx 500)$ than in comparable studies of the living tetramer $(K_{T_1}/K_\infty \approx 6)$. Moreover, the relative accuracy of M_e determinations diminishes as M_0 decreases, whilst studies in the low range of M_0 are crucial for the determination of K_1. Exceptional care is therefore essential in these experiments, details of which are given in reference (13).

For $K_1 > K_\infty$, the equation of Tobolsky gives the apparent constant K'_1 smaller than K_1 and, hence, extrapolation proceeds through a set of points lying *below* the value of K_1. In testing the reliability of extrapolation, it is desirable to use an alternative method of calculation giving the apparent K''_1 greater than K_1 and thus, one may approach K_1 from above. This was done by VRANCKEN et al. (13) for the region $M_0/P^*_{total} < 1$ by assuming $K_2 = K_3 \cdots = K_\infty = 0$. This leads to the equation

$$K''_1 = (M_0/M_e - 1)/\{P^*_{total} - (M_0 - M_e)\} \text{ leading to } K''_1 > K_1 \text{ for any } M_0.$$

The limiting value K_1 is given again by $\lim K''_1$ for $M_0/P^*_{total} \to 0$ but extrapolation involves now a set of points greater than K_1. Using this approach it was possible to obtain a reliable estimate of the accuracy for each determination of K_1, see, e.g., reference (13). This discussion shows again the futility of calculating K_3 and exposes the great uncertainties involved in the calculation of K_2. Although VRANCKEN et al. attempted in their preliminary publication (22) to determine K_2 for the T_1 tetramer system, later work indicated that the claimed values are doubtful. It seems, nevertheless, that for that system K_2 may be slightly greater than K_∞, while K_1 is about six times as large. Kinetic work, described later, indicates that the probable increase in the K_2 value, when compared with K_∞, arises from a smaller depropagation rate constant, whereas the propagation rate constant, $T_1\alpha + \alpha \to T_1\alpha\alpha$ differs insignificantly from that observed for a high-molecular weight polymer.

Finally, it should be stressed that the nomenclature used in this section refers to consecutive additions of monomers to the same end. It is assumed implicitly that the equilibrium constants for the reactions $\alpha T_1 + \alpha \rightleftarrows \alpha T_1\alpha$ and $T_1 + \alpha \rightleftarrows T_1\alpha$ are virtually identical, and by defining P^*_{total} as the total concentration of *livings* ends, one incorporates this assumption into the calculations.

6. Kinetic techniques for the determination of the equilibrium constant of propagation

The equilibrium constant of any process is related to the rate constants of the forward and backward reactions, namely,

$$K = k_{forward}/k_{backward} .$$

Hence, the equilibrium propagation constant may be calculated, if the absolute propagation and depropagation rate constants are determined. Various techniques used in the determination of the absolute rate constants of anionic propagation and depropagation were discussed elsewhere (53). At this stage, only one kinetic method will be described which simultaneously gives the propagation and depropagation rate constants as well as the equilibrium constant. The method is illustrated by the reaction, whereby upon addition of α-methylstyrene to the living α-methylstyrene dimer, $^-\alpha\alpha^-$, the latter yields reversibly the respective living α-methylstyrene trimer, i.e.

$$^-\alpha\alpha^- + \alpha \underset{k_{-1}}{\overset{k_1}{\rightleftarrows}} {}^-\alpha\alpha\alpha^- \dots K_D .$$

The system, $^-\alpha\alpha^- + \alpha \rightleftarrows {}^-\alpha\alpha\alpha^-$, was investigated by LEE, SMID, and SZWARC (23) who studied its kinetics by using a stirred-flow reactor. The great simplicity of this device arises from the fact that the concentrations of all the reagents, products, and intermediates remain constant in the reactor (24). The reagents are fed into the reactor at a constant rate, say v cc/second. A solution containing reagent A at concentration $C_{A,0}$, B at $C_{B,0}$, etc. flows in and simultaneously v cc/second of a solution-containing the products flow out. Efficient stirring in the reactor main tains the composition of the reacting solution homogeneous throughout its entire volume, V, and, hence, the composition of the outflowing liquid is identical with that prevailing in the reactor. In the living dimer → living trimer system, the reaction is quenched instantly in the outgoing liquid by leading it into wet tetrahydrofuran.

The simplicity of such a system is evident from the fact that one needs only to determine, for different rates of flow, the concentration of the residual monomer in the quenched reaction fluid in order to get all the required constants. The initial concentration of the monomer, x_0, and that of the living dimer, y_0, are chosen at will and kept constant in each series of experiments. The reactor is operated for a few minutes to allow the system to reach its stationary state and then a sample of the outgoing liquid is withdrawn and analyzed for α-methylstyrene. The result gives, therefore, the stationary concentration of the monomer, x_t, in the reactor.

The stationary concentrations of the dimer, y_t, and of the trimer[1], z_t, are then determined by the stoichiometry of the process, i.e.

$$y_t = y_0 - (x_0 - x_t) \text{ and } z_t = (x_0 - x_t) .$$

The amount of a reagent consumed in the reactor is given by the mass balance, i.e. by the difference between the amounts flowing in and out, thus enabling the calculation of the required rate constants. For monomer, the mass balance becomes:

$$v (x_0 - x_t) = V \{2 k_1 x_t y_t - k_{-1} z_t\}$$

or

$$v (x_0 - x_t) = V \{2 k_1 x_t [y_0 - (x_0 - x_t)] - k_{-1} (x_0 - x_t)\} .$$

The factor 2 arises from the presence of two living ends in each dimer, both of them being capable of reaction with monomer. If t — the residence time of the liquid in the reactor — is defined as the ratio V/v, one arrives at the following expression

$$1/t = 2 k_1 x_t \{y_0/(x_0 - x_t) - 1\} - k_{-1} .$$

Hence, a plot of $1/t$ against $x_t \{y_0/(x_0 - x_t) - 1\}$ should give a straight line as shown in Fig. 9, with a slope equal to $2 k_1$. The intercept on the

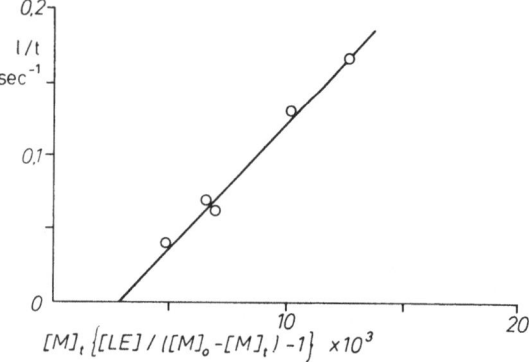

Fig. 9. Graphic solution of the balance equation for the reaction $^-\alpha\alpha^- + \alpha \rightleftarrows {}^-\alpha\alpha\alpha^-$. α denotes α-methylstyrene; $^-\alpha\alpha^-$ its living dimer; and $^-\alpha\alpha\alpha^-$ its living trimer. $1/t$ is the reciprocal of the resident time in the stirred-flow reactor. Reproduced, with permission, from Lee, Smid, and Szwarc:J. Am. Chem. Soc. **85**, 912 (1963).

$1/t$ axis is equal to $-k_{-1}$ while that on the other axis is related to the equilibrium constant, i.e. $K_D = (1/2) \text{(intercept)}^{-1}$. These constants, obtained for different initial conditions of the reaction, are given in Table 5, and their self-consistency demonstrates the reliability of the method.

---

[1] Note that z_t gives the concentration of α-methylstyrene units added to $^-\alpha\alpha^-$, whether they form the trimer or the tetramer. It is implicitly assumed that the added unit has the same chance to dissociate whatever the state of the other end of the dimer.

Addition of 2 or more monomer units to a particular end of the dimer was ignored in deriving the above equation. This is justified because the rate of addition to the right end of the trimer,

$$^-C(CH_3)(Ph).CH_2.CH_2.C(CH_3)(Ph).CH_2.C(CH_3)(Ph)^-,$$

was found to be much slower than that to the left end. The latter reaction is accounted for in the equation by introducing the factor $2 k_1$ instead of k_1. In this procedure it is implied that the rate of the reaction at one end of the dimer is not affected by the fate of the other end, an assumption which seems justified by the data.

The kinetic method produced results concordant with those derived from the direct static approach (13), and leads to a value of 331 liter/mole for K_D as compared with $K_D = 240 \pm 30$ liter/mole obtained by extrapolation of K'_D to $M_0 = 0$. In view of the steepness of the extrapolation, the kinetic value seems to be more reliable.

Table 5. Constants for the reversible reaction $^-\alpha\alpha^- + \alpha \overset{k_1}{\underset{k_{-1}}{\rightleftarrows}} {}^-\alpha\alpha\alpha^- \ldots K_D^1$.

Solvent = Tetrahydrofuran. $T = 25°$ C. Conter ion = K^+

P^*_{total} $M \times 10^3$	M_0, $M \times 10^3$	k_1, Liter/Mole.Second	k_{-1}, Second^{-1}	K_D, Liter/Mole
3.0	3.0	17.2	0.050	346
4.4	2.9	17.3	0.052	333
4.5	1.1	16.7	0.049	339
6.5	2.9	18.9	0.062	304
14.0	3.0	15.4	0.046	333
	Average:	17.1	0.052	331

[1] Data of LEE, SMID, and SZWARC (23).

7. Factors affecting the equilibrium constant of propagation

Steric strain in the polymer and the rigidity of its chain are the two most important factors affecting the equilibrium propagation constant. Bulky substituents on the C=C group of a vinyl monomer are the main cause of steric strain in its polymer and the hindrance becomes considerable when two substituents are attached to the same carbon atom. The strain is manifested by a relatively low heat of polymerization, e.g., $-\Delta H$ for ethylene polymerization is ~ 24 kcal./mole, for styrene it amounts to $17-18$ kcal./mole, while polymerization of α-methylstyrene evolves only $7.5-8$ kcal./mole. Inspection of models shows that substituents on alternate carbon atoms of a $-C-C-C-$ chain interfere with each other to a greater extent than those located on adjacent atoms. For example, a direct linkage of two tert.-butyl groups gives a relatively strainless hydrocarbon, namely, 2,2,3,3-tetramethylbutane

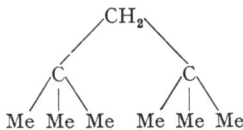

whereas their linkage through a CH_2 group leads to severe strain.

$$\begin{array}{c} CH_2 \\ / \quad \backslash \\ C \qquad C \\ /|\backslash \quad /|\backslash \\ Me\ Me\ Me \quad Me\ Me\ Me \end{array}$$

These considerations explain why a head-to-head — tail-to-tail polyiso-butene $—CMe_2.CH_2.CH_2.CMe_2.CMe_2.CH_2.CH_2.CMe_2—$ is expected to be much more stable than the conventional head-to-tail-polyisobutene, $—CH_2.CMe_2.CH_2.CMe_2—$. The same reasons partially explain the stability of the living α-methylstyrene tetramer, T_1, $^-C(Me)(Ph).CH_2.CH_2.$ $.C(Me)(Ph).C(Me)(Ph).CH_2.CH_2.C(Me)(Ph)^-$ which, as remarked previously, does not decompose into lower oligomers, whereas the living tetramer, T_2, $^-C(Me)(Ph).CH_2.C(Me)(Ph).CH_2.CH_2.C(Me)(Ph).CH_2.$ $.C(Me)(Ph)^-$ exists only in equilibrium with the monomer and the respective trimer and dimer. The same factors account for the lower heat of polymerization of α,α-substituted vinylidene monomers as compared with that of analogous α-β-substituted olefins.

Steric strain is maximal when the addition of monomer causes the substituents of the preceding unit to become crowded by bulky groups on *both* sides. For example, in the conversion of living α-methylstyrene dimer into trimer

$$^-C(Me)(Ph).CH_2.CH_2.C^-(Me)(Ph) + CH_2:C(Me)(Ph) \rightleftarrows$$
$$^-C(Me)(Ph).CH_2.CH_2.C(Me)(Ph).CH_2.C(Me)(Ph)^- \ldots K_{D,1}$$
$$\quad 1 \qquad\qquad 2 \quad 3 \quad 4 \qquad\quad 5 \quad 6$$

the strain between the substituents on carbon atoms 4 and 6 may be reduced by a conformational adjustment of the substituents on carbon 4. Subsequent addition of another monomer unit, viz.,

$$^-C(Me)(Ph).CH_2.CH_2.C(Me)(Ph).CH_2.C(Me)(Ph)^- + CH_2:C(Me)(Ph) \rightleftarrows$$
$$^-C(Me)(Ph).CH_2.CH_2.C(Me)(Ph).CH_2.C(Me)(Ph).CH_2.C(Me)(Ph)^- \ldots K_{D,2}$$
$$\quad 1 \qquad\quad 2 \quad 3 \quad 4 \qquad\quad 5 \quad 6 \qquad\quad 7 \quad 8$$

introduces strain between substituents on carbon atoms 6 and 8 which cannot be released, because the groups on carbon 6 are now squeezed between those located on carbons 4 and 8. The same type of strain results upon each subsequent monomer addition. One expects, therefore, the equilibrium constant $K_{D,1}$ to exceed $K_{D,2}$ and the latter should be similar in magnitude to K_∞. Studies by VRANCKEN et al. (13) confirmed this supposition. However, the expected decrease in the heat of polymerization was not detected. On the other hand, a direct calorimetric study of the system indicated a substantial increase in the heat of the reaction $^-\alpha\alpha^- + \alpha \rightleftarrows {}^-\alpha\alpha\alpha^-$ when compared with that evolved in the polymerization (54).

Electrostatic repulsion between the two charged ends (or the respective C^-, K^+ dipoles) of the living α-methylstyrene dimer may also contribute to its heat of polymerization, because the electrostatic energy of the system decreases as $^-\alpha\alpha^-$ grows to $^-\alpha\alpha\alpha^-$. On these grounds, one expects the equilibrium constant $K_{D,1}$ to be larger than the analogous $K_{T_1,1}$ constant,

$^-$C(Me) (Ph) . CH$_2$. CH$_2$. C(Me) (Ph) . C(Me) (Ph) . CH$_2$. CH$_2$. C(Me) (Ph)$^-$ +
   + CH$_2$: C(Me) (Ph) $\leftrightarrows$

$^-$C(Me) (Ph) . CH$_2$. CH$_2$. C(Me) (Ph) . C(Me) (Ph) . CH$_2$. CH$_2$. C(Me) (Ph) .
   . CH$_2$. C(Me) (Ph)$^-$ . . . $K_{T_1,1}$.

This indeed was found by VRANCKEN et al. (13), although here again the expected decrease in the heat of polymerization was not observed. The importance of the electrostatic effect could be tested by comparing the equilibrium constant $K_{D,1}$ ($^-\alpha\alpha^- + \alpha = {}^-\alpha\alpha\alpha^-$) with that of the cumylpotassium + α-methylstyrene process which gives the relevant dimer containing one living end only. Since no electrostatic contribution is associated with the free energy of the latter reaction, its heat might be expected to be lower than that of the former. Preliminary studies of the cumylpotassium system have been carried out and, surprisingly, the equilibrium constant of this reaction was found to be even higher than K_D. One must conclude, therefore, that electrostatic repulsion, if any, results from the interaction of dipoles — not of free charges — and apparently its contribution to the free energy of the addition reaction seems relatively small.

Steric strain may also decrease the entropy of the polymer chain by increasing its rigidity and, hence, it could increase the value of $-\Delta S_p$. However, the observed changes in the entropy of polymerization are relatively insignificant, e.g., $-\Delta S_p$ for styrene polymerization to a solid polymer amounted to ~ 25 e.u., whereas the corresponding value for α-methylstyrene was found to be ~ 26 e.u. This is indeed a small change when compared with a 10-kcal./mole decrease in the heat of polymerization. The growth of oligomers should probably lead to a smaller change in the entropy of the system than the growth of analogous high polymers, and this factor may partially account for their higher equilibrium constants.

Finally, it should be stressed that, in contradistinction to high-molecular weight polymers, the equilibrium growth constants of low-molecular weight oligomers depend on the mechanism of the reaction and on the nature of the reactive end groups. This point was recognized and stressed by Dainton and Ivin (2).

8. Molecular weight distribution in monomer — living polymer equilibria

In discussing the equilibria between a monomer and its living polymer, two aspects of the problem should be considered: (1) establishment of an equilibrium, or more correctly of a stationary state, between monomer and growing ends; (2) establishment of an equilibrium between *all* the growing polymers. The first process approaches its equilibrium state fairly rapidly, whereas the other is usually very slow and requires a long time to produce the ultimate equilibrium distribution.

It was shown (25) that for a solution of living polymers, the ultimate equilibrium molecular weight distribution is of the Flory type, (26 a) i.e., the mole fraction of the $(n_0 + j)$-mer is given by $(K \cdot M_e)^j \cdot (1-KM_e)$, if all the equilibrium constants are identical. In accordance with our symbolism, $P_{n_0}^*$ defines the lowest living n_0-mer which may grow but not degrade, and its mole fraction in the equilibrated mixture should be $(1-KM_e)$. For a high average degree of polymerization $1-KM_e \ll 1$ and, therefore, only a minute fraction of polymers exists as n_0-mers.

Whenever a high-molecular weight polymer is produced, the propagation rate constant, k_p, must be much greater than k_d, the depropagation constant; the initial concentration of the monomer, M_0 greater than M_e-its equilibrium concentration, and $M_0/(P_{n_0}^*)_0 \gg 1$. Polymerization in such a system usually leads to a Poisson molecular weight distribution (26 b) when nearly all the polymeric molecules have a degree of polymerization close to $M_0/(P_{n_0}^*)_0$. Although such a system is not yet in its true equilibrium state, the polymerization is essentially completed, and the concentration of the monomer differs only insignificantly from that attained at the ultimate equilibrium. In fact, its value, M_e, is determined approximately by the condition of the stationary state, i.e.

$$k_p M_e' \left\{ \sum_0^\infty P_{n_0+j}^* \right\} = k_d \left\{ \sum_1^\infty P_{n_0+j}^* \right\}$$

leading to $M_e' = (1/K)(1 - P_{n_0,t}^*/P_{total}^*)$ which is only insignificantly greater than M_e, because $P_{n_0,t}^*$ is expected to be only slightly smaller than $P_{n_0,eq}^*$. A numerical example shows how small is the difference between M_e' and M_e. Let us consider a system where initial monomer concentration is 1 mole/l, total concentration of living polymers equals 0.01 mole/l, and the propagation equilibrium constant $K = 99$ l/mole. The calculated ultimate $M_e = 0.01$ M, corresponding to $P^*_{n_0, eq} = 10^{-4}$ M. For the Poisson distribution $P^*_{n_0, t} = 0$ and $M'_e = 0.0101$ mole/l, i.e. only 1% higher than the ultimate M_e.

The redistribution of molecular weight in the living polymer system represents an interesting kinetic problem. In this reaction, neither the concentration of the polymerized material nor its number average molecular weight changes. Hence, had the process been studied in a dilatometer or in an osmotic cell, no change would be observed. However, the weight average molecular weight, and all the higher average molecular weights, increase as the distribution changes from the Poisson to the Flory type. The differential equation giving $d(\overline{DP}_w)/dt$ was discussed by BROWN and SZWARC (25). Assuming the steady state approximation, they found

$$d(\overline{DP}_w)/dt = 2 k_d/(\overline{DP}_n) \, [1 - x_0(\overline{DP}_n)]$$

where $(\overline{DP}_w)$ and $(\overline{DP}_n)$ denote the respective weight average and number average degree of polymerization, k_d is the depropagation rate constant, and x_0 the mole fraction of the lowest living n_0-mer. A more rigorous treatment leads to

$$d(\overline{DP}_w)/dt = k_d \{[(\overline{DP}_w) - 1]/(\overline{DP}_n)\} \cdot (x_{0,eq} - x_0)$$

or

$$d(\overline{DP}_w)/dt \simeq k_d \{(\overline{DP}_w)/(\overline{DP}_n)\} \cdot (x_{0,eq} - x_0)$$

where $x_{0,eq}$ is the mole fraction of the living n_0-mer at the ultimate equilibrium state. Since $x_{0,eq} = 1/(\overline{DP}_n)$, the latter equation becomes identical with the former if $(\overline{DP}_w)/(\overline{DP}_n) = 2$, i.e. for a system being very near to its ultimate equilibrium state. These equations make it obvious that the redistribution of polymer chains is an extremely slow process at a high degree of polymerization, because the rate is inversely proportional to $\overline{DP}_n$.

The redistribution of molecular weights, discussed above, results from reactions taking place at the polymeric *ends* only, viz.,

$$P^*_n + M \rightleftarrows P^*_{n+1} \, .$$

However, in some systems the redistribution of monomer segments may result from other reactions, such as trans-esterification, trans-amidation or trans-etherification. Consider, e.g., a solution of living polyglycols

$$\text{----}CH_2CH_2O.CH_2.CH_2O^-, \, Na^+ \, .$$

As the result of a reaction,

$$\overset{l \text{ units}}{\overbrace{\text{----}^-CH_2.CH_2.O}}.\overset{k \text{ units}}{\overbrace{CH_2.CH_2.O\text{----}}} +$$

$$\overset{j \text{ units}}{\overbrace{^-OCH_2.CH_2\text{----}}} \rightarrow$$

$$\overset{l \text{ units}}{\overbrace{\text{----}CH_2.CH_2.O^-}} + \overset{j + k \text{ units}}{\overbrace{\text{----}CH_2.CH_2.O.CH_2.CH_2.O\text{----}}}$$

an $(l + k)$-mer and a j-mer are converted into an l-mer and a $(j + k)$-mer. The ultimate equilibrium distribution resulting from this reaction is again of a Flory type; however, the kinetics of this process is different from those previously considered. Its mathematical treatment has been reported recently by Hermans (27).

9. Sharpness of transition phenomena in equilibrium polymerization

In his approach to the thermodynamics of propagation, Dainton was concerned with a process transforming free monomer into monomer segments of a *high*-molecular weight polymer. The free energy change of this reaction is given by the equation

$$\Delta F_p = \Delta F(M_s) - \Delta F(M_f),$$

where the molar free energy of the monomer, $\Delta F(M_f)$, involves the concentration term $RT\ln(M_f)$, whereas no equivalent term appears in the free energy, $\Delta F(M_s)$, of the monomer segments, because the segments are part of already existing macromolecules. This gives to the process the characteristic feature of a physical aggregation such as, e.g., crystallization. The concentration of polymer affects ΔF_p, indirectly, only in as much as it determines the deviation of the monomer's activity from its ideal value by influencing the monomer-solvent-segment and segment-segment interactions. In this terminology, the ceiling temperature has a uniquely defined value determined by the equation $\Delta F_p = 0$ for constant thermodynamic conditions of the system, i.e. for chosen concentrations of monomer and polymer and for the particular nature of the solvent.

Alternatively, we may look at the state of ultimate equilibrium of a system containing a monomer, the polymerization of which involves no termination and yields living polymers. In such a discussion, the modes of initiation of the polymerization and the concentration of the initiator or of the lowest living oligomer must be specified.

This approach has been developed by Tobolsky (9, 28), who considered three distinct systems differing in their modes of initiation, viz.,

I. $\quad I + M \rightleftarrows IM^* \quad \ldots\ldots\ldots\ldots K_1$
$IM^* + M \rightleftarrows IM_2^*$
$\ldots\ldots\ldots\ldots\ldots\ldots\} \ldots\ldots\ldots\ldots K$
$IM_{n-1}^* + M \rightleftarrows IM_n^*$

II. $\quad M \rightleftarrows M^* \quad \ldots\ldots\ldots\ldots K_1$
$M^* + M \rightleftarrows M_2^*$
$\ldots\ldots\ldots\ldots\ldots\ldots\} \ldots\ldots\ldots\ldots K$
$M_{n-1}^* + M \rightleftarrows M_n^*$

III. $M + M \rightleftarrows M_2^*$ .... K_1

$$M_2^* + M \rightleftarrows M_3^*$$
$$.................$$ .... K
$$M_{n-1}^* + M \rightleftarrows M_n^*$$

A system containing a living n_0-mer which may grow but not degrade is exemplified by Case I. Polymerization of cyclic monomers initiated by spontaneous ring opening provides an example of Case II. Finally, a hypothetical thermal polymerization of a vinyl monomer arising from bimolecular initiation and proceeding without termination illustrates Case III.

We may limit the discussion to Case I without losing sight of the generality of behavior which characterizes all three processes. The system is determined by the initial concentrations of monomer, M_0, initiator, P_0^*, and the equilibrium constants K_1 and K. Its qualitative behavior is not affected by the simplifying assumption demanding K_1 to be equal to K and therefore, for the sake of clarity, this is accepted in the present discussion.

The state of equilibrium is characterized by the equilibrium concentration of the monomer, M_e, and the number average degree of polymerization, $\bar{j}$, the latter being defined by the equation $(\bar{j}) = \sum\limits_1^\infty j \cdot P_j^* / \sum\limits_1^\infty P_j^*$.
M_0, P_0, M_e, and $\bar{j}$ are related by the following equations derived from first principles, viz.,

$$(\bar{j}) = (M_0 - M_e)/(P_0^* - P_{0,e1}^*) = 1/(1 - KM_e)$$
$$P_0^* = (\bar{j}) P_{0,e1}^*$$
and
$$(\bar{j})^2 KP_0^* + (\bar{j})[K(M_0 + P_0^*) - 1] - 1 = 0 .$$

The equilibrium constant K is a function of temperature, T; hence, the equilibrium concentration of monomer, M_e, the degree of polymerization, $\bar{j}$, and the fraction of polymerized monomer, $(M_0 - M_e)/M_0$, are all unique functions of T for constant M_0 and P_0^*. For the α-methylstyrene – living poly – α-methylstyrene system, these functions are shown in Figs. 10 and 11. The steep decrease in j or in $(M_0 - M_e)/M_0$ within a narrow temperature range is the striking feature of these curves. This behavior characterizes all equilibrium polymerizations and the sudden increase in $\bar{j}$ on lowering the temperature shows the similarity between these phenomena and other transitions based on physical aggregation.

For equilibrium polymerizations, the sharpness of transition has a well-defined physical meaning and may be defined mathematically by, e.g., $d\bar{\jmath}/dT$ in the transition region. This derivative may be evaluated by differentiating the quadratic equation for $\bar{\jmath}$ with respect to T and

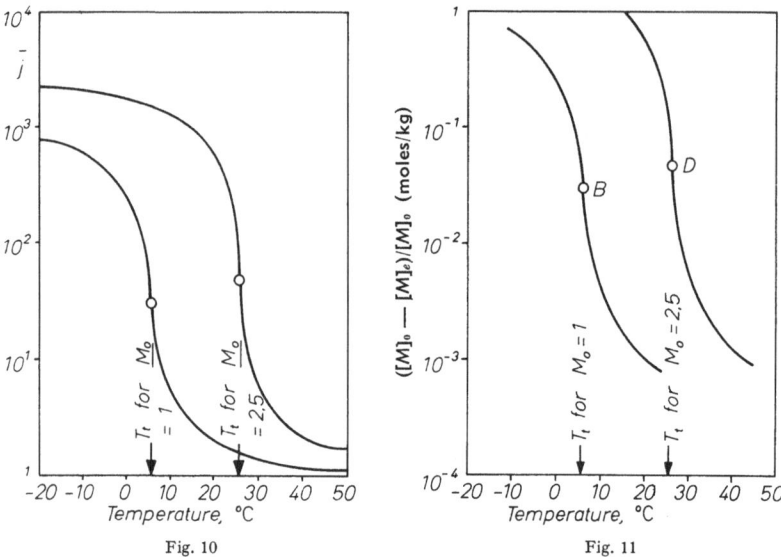

Fig. 10   Fig. 11

Fig. 10. Degree of polymerization, $\bar{\jmath}$, versus temperature in °C for equilibrium polymerization of α-methylstyrene. A. initial concentration of the nonomer 1 mole/kg. B. initial concentration of the monomer 2,5 moles/kg. Concentration of living ends 0,001 moles/kg. Reproduced, with permission, from Tobolsky and Eisenberg: J. Colloid Sci. **17**, 49 (1962).

Fig. 11. Amount of polymerized monomer, $[M]_e = (M_0\text{-}M)$ moles/kg versus temperature for equilibrium polymerization of α-methylstyrene. $M_0 = 1$ mole/kg or $M_0 = 2,5$ moles/kg. Concentration of living ends 0,001 moles/kg. Reproduced, with permission, from Tobolsky and Eisenberg: J. Colloid Sci. **17**, 49 (1962).

substituting $(d\bar{\jmath}/dK)_T \cdot (dK/dT)$ for $d\bar{\jmath}/dT$. At the transition temperature, the result may be set in the approximate form

$$d\bar{\jmath}/dT = (\Delta H_p/\mathrm{RT}^2)/K P_0^* [2 - (K P_0^*)^{1/2}]$$

by using the simplifying relations

$$\bar{\jmath}_{\text{transition}} \approx (K P_0^*)^{-1/2}$$

and

$$(M_0 - M_e)_{\text{transition}} \approx (P_0^*/K)^{1/2} .$$

The latter are plausible approximations because at the transition temperature, and above it, the initial monomer concentration, M_0, is nearly equal to its equilibrium concentration, i.e. $M_0 \approx M_e \approx 1/K$. In view of the last relation, the ceiling temperature of the system is defined as the temperature at which $K(T) = 1/M_0$.

It is interesting to compare Dainton's definition of the ceiling temperature with that of Tobolsky. Both have the same mathematical form. Dainton's definition refers to the single process: free monomer $\rightarrow$ $\rightarrow$ monomer segment of a *high* polymer, and takes no account of the building process leading to a macromolecule. On the other hand, TOBOLSKY refers to the whole stepwise building reaction and gives full consideration to the initiation processes. Therefore, the sharpness of the transition which is irrelevant in DAINTON's definition is pertinent in TOBOLSKY's treatment. However, DAINTON's definition is applicable to systems in which equilibrium between various living polymers has not yet been established, e.g., the photodegraded systems studied by IVIN or the pseudo-equilibrium situation when a Poisson distribution is attained as a consequence of rapid polymerization of living polymers. On the other hand, TOBOLSKY's definition applies only when a complete equilibrium is established between *all* the polymeric molecules as well as between the polymer and the monomer.

Finally, the factors affecting the sharpness of the transition should be examined. The expression giving $[d\bar{\jmath}/DT]_{P_0^* M_0}$ shows clearly that the transition is sharper, the greater ΔH_p (the heat of polymerization) and the smaller P_0^* (the number of initiating species). The effect of ΔH_p is obvious: increase in its absolute value narrows the temperature range in which K varies from values much larger than $1/M_0$ to those much smaller. The effect of P_0^* is equally easy to comprehend. The amount of polymerized monomer increases on lowering the temperature of the system. For lower P_0^*, the latter is distributed among fewer chains, giving a greater increase in $\bar{\jmath}$.

In a similar way, one may calculate the degree of polymerization, $\bar{\jmath}$, as a function of M_0 for constant temperature and constant P_0^*. The results are presented in Fig. 12 and show again a sharp transition occurring within a critical region of M_0. The sharpness of the transition is defined by $(d\bar{\jmath}/dM_0)_T$ in the vicinity of the critical concentration M_0 and is given approximately by the equation $[d\bar{\jmath}/dM_0]_T = 1/2\, P_0^*$.

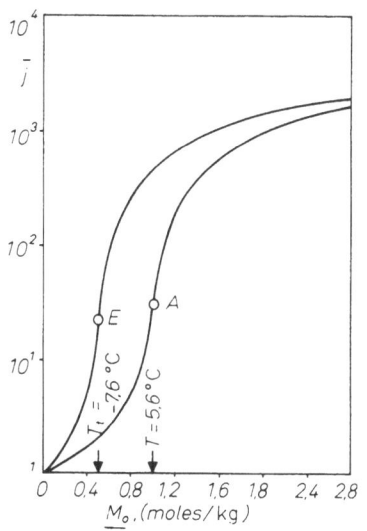

Fig. 12. Degree of polymerization, $\bar{\jmath}$, versus initial monomer concentration M_0 of α-methylstyrene. Equilibrium polymerization at +5,6° C or -7,6° C. Concentration of living ends 0,001 mole/kg. Reproduced, with permission, from TOBOLSKY and EISENBERG: J. Colloid. Sci. **17**, 49 (1962).

The greater sharpness of the ceiling temperature arising from an increase in the size of the polymer has its analogy in the behavior of the melting point of a crystal (2). The melting temperature is sharp for large crystals but it becomes a broad range for small crystallites because the contributions of the surface, edges, and apexes to the free energy of the system become significant when compared with the bulk free energy. The similarity to the effect due to the presence of polymer chain ends is obvious.

These effects were observed in experiments reported by DAINTON's group. For example, they were clearly demonstrated in the study of the cationic polymerization of α-methylstyrene (29). The formation of low-molecular weight polymers above the ceiling temperature was attributed to a higher heat of polymerization when oligomers are formed. All of the kinetic studies by DAINTON's group showed that the rate of polymerization does not drop abruptly to zero at the ceiling temperature but that in its vicinity the curve, giving rate as a function of temperature, becomes asymptotic to the T axis. This again points to a broadening of the transition arising from a decrease in the molecular weight of the resulting polymer.

Finally, the data published by GEE (30) permit one to evaluate the sharpness of a transition involving floor temperature. Gee studied the temperature dependence of the viscosity of liquid sulfur and observed its sudden, steep increase at a critical temperature followed by its decrease at still higher temperatures. He developed the first, relatively complete theory of equilibrium polymerization of liquid sulfur (30) from which he estimated the chain length of the polymeric sulfur at various temperatures. His results have been recently confirmed by experimental measurements of magnetic susceptibility of the liquid sulphur (50) and its electron spin resonance (51).

GEE's theory was unified by TOBOLSKY and EISENBERG (31) and further improved by TOBOLSKY et al. (52) who removed the original restriction demanding an multiple numbers of 8 S atoms in each chain. This theory demands the existence of a transition (floor) temperature and excellently accounts for its sharpness as determined from GEE's results. A similar situation is observed in liquid selenium (32).

10. Deviations from the law of ideal solutions

It has been remarked in the preceding sections that the equilibrium concentration of monomer in solution of its living polymer is affected by the nature of the solvent and by the polymer concentration, because these factors influence the activities of the components. A quantitative treatment of these effects, based on SCOTT's modification of the standard lattice theory of polymer solutions (33), has been outlined recently by BYWATER (34).

The polymerization of a mole of a liquid monomer to form a solid polymer is associated with a decrease in the free energy of the system given by the equation

$$- \Delta F_{\text{liquid, solid}} = RT\{\ln a_p - \ln a_m\}$$

where a_m and a_p denote, respectively, the activities of the free monomer and of the monomer segments in the polymer chain when they are at equilibrium with each other in the investigated solution. The activities refer to liquid monomer and solid polymer as their respective standard states. The lattice theory of polymer solutions gives these activities in terms of the volume fractions of the components, i.e. ϕ_s, ϕ_m, and ϕ_p, and of the three interaction parameters, $\mu_{s,m}$, $\mu_{s,p}$, and $\mu_{m,p}$ of the respective pairs: solvent-monomer, solvent-polymer, and monomer-polymer. Thus,

$$-\Delta F_{\text{liq, solid}}/\text{RT} = (\mu_{s,p} - \mu_{s,m})\,\phi_s - 1 - \ln\phi_m + \mu_{m,p}(\phi_m - \phi_p)$$

which may be converted into ΔF_p (the free energy of polymerization of a 1 M. monomer solution to a 1 M. solution of polymeric segments) by adding to the equation the respective free energy of 1 M. solution of the monomer and of the polymeric segments, i.e.

$$\Delta F_m^*/\text{RT} = \mu_{s,m}\phi_s^* + \phi_s^*(1 - \phi_s^*)^{-1}\ln\phi_s^* + \ln\phi_m^*$$

and

$$\Delta F_p^*/\text{RT} = \mu_{s,p}\phi_s^* - 1 + \phi_s^*(1 - \phi_s^*)^{-1}\ln\phi_s^*\,,$$

ϕ^* denoting the respective volume fractions of the monomer, solvent, and polymer in the relevant 1 M. solutions. This leads to the equation

$$\ln K = -\Delta F_p/\text{RT} = \ln(\phi_m^*/\phi_m) + (\mu_{s,p} - \mu_{s,m})(\phi_s - \phi_s^*) + \mu_{m,p}(\phi_m - \phi_p)$$

giving the equilibrium propagation constant in terms of the mole fractions of the components and the relevant interaction parameters.

When $\mu_{s,p} = \mu_{s,m}$, and if either $\mu_{m,p} = 0$ or ϕ_m and ϕ_p tend to zero, the equation for $\ln K$ reduces to $\ln K = -\ln(\phi_m/\phi_m^*)$. Moreover, if $\phi_m/\phi_m^* = [M]_e$, as would be expected for such a system, the laws of ideal solution apply, i.e. $\ln K = -\ln[M]_e$ or $K[M]_e = 1$.

For experiments carried out in a highly dilute solution of living polymer, $\ln K$ is given by the simplified equation

$$\ln K = \ln(\phi_m^*/\phi_m) + (\mu_{s,p} - \mu_{s,m})(\phi_s - \phi_s^*) + \mu_{m,p}\phi_m\,.$$

Deviations from ideal behavior arise from three causes:

(1) ϕ_m is not proportional to the monomer concentration in which event $\phi_m/\phi_m^* \neq [M]_e$.

(2) The interaction of solvent with monomer is different from that with polymer, i.e. $\mu_{s,m} \neq \mu_{s,p}$.

(3) The monomer-polymer interaction is not negligible, i.e. $\mu_{m,p} \neq 0$.

The deviations due to variation of the partial volume of monomer with dilution are expected to be negligible in not too concentrated solutions. For example, if the partial volume of the monomer remains constant up to 1 M. solution and $[M]_e < 1$ mole/l, then $\phi_m/\phi_m^* = [M]_e$.

The inequality $\mu_{s,m} \neq \mu_{s,p}$ accounts for the dependence of $[M]_e$ on the nature of the solvent. Its magnitude is shown, e.g., by the results of Bywater and Worsfold (*14*) who found the equilibrium concentration of styrene to be different in hexane and in benzene solutions (see p. 465 and Fig. 3). The effect of solvent is not eliminated even at an extremely low equilibrium monomer concentration; in fact, in such a case, the deviation from ideal behavior is given by $(\mu_{s,m} - \mu_{s,p})(1 - \phi_s^*)$.

The effect of solvent may be separated into heat and entropy terms, the former given by the difference in the molar heats of dilution of the monomer and polymer solutions from the standard solvent volume fractions ϕ_s^* to their equilibrium fractions ϕ_s. The entropy term is constant

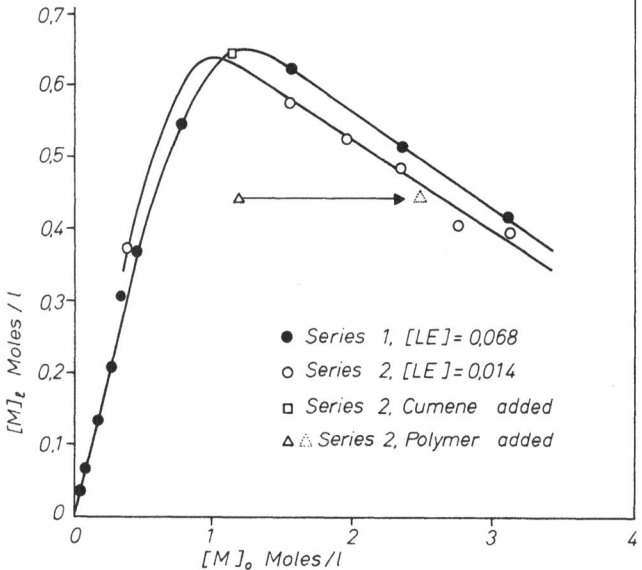

Fig. 13. Equilibrium concentration, $[M]_e$, of α-methylstyrene in contact with its living polymer in THF at 0° C, plotted as a function of its initial concentration $[M]_e$. Reproduced, with permission, from Vrancken, Smid, and Szwarc; Trans. Faraday Soc. **58**, 2036 (1962).

for many solvents. However, it becomes of great importance when the random configuration of the polymer chain, assumed to prevail in "normal" solutions, is drastically changed by the solvent, e.g., when a polymer acquires a helical structure in the investigated solution.

Finally, the last term, $\mu_{m,p}(\phi_m - \phi_p)$ is unimportant if ϕ_m and ϕ_p are small. However, it becomes significant in concentrated polymer solution as shown, e.g., by the studies of Vrancken et al. (*13*). Their investigation of the α-methylstyrene − living poly-α-methylstyrene system in tetrahydrofuran solution demonstrated that for a *high* concentration of living ends the equilibrium concentration of the monomer increases initially and thereafter decreases as more and more monomer is added to the system. This strange behavior, illustrated by Fig. 13, contrasts with that shown in Fig. 7 where, for low concentrations of living ends, $[M]_e$ increases monotonically with $[M]_0$ and eventually reaches a limiting value characteristic of an equilibrium involving a high-molecular weight living polymer. The decrease in M_e shown in Fig. 13 results from an increase in the *total* concentration of polymer. For example, upon the addition of dead poly-α-methylstyrene to a solution of living poly-α-methylstyrene which was in equilibrium with its monomer, the equilibrium concentration of the latter descreased from 0.63 mole/l (the square point in Fig. 13) to about 0.42 mole/l (the triangle point in the same figure). By adding the mass of the dead polymer to $[M]_0$, the effective $[M]_0$ is calculated and this, as shown by the dotted triangle, corresponds to the anticipated value of $[M]_e$. In terms of the derived equation, the effect arises from an increase in the term $\mu_{m,\,\flat}\cdot\phi_\flat$ caused by the addition of polymer, or of monomer which eventually polymerized. The addition of cumene, on the other hand, has a negligible influence on the $[M]_e$ value (*13*).

The decrease in $[M]_e$ upon addition of monomer superficially seems to contradict the law of mass action. However, addition of monomer increases its *activity* in the equilibrated solution, although the increase becomes negligible as the activity approaches its limiting value. On the other hand, the resulting increase in polymer *concentration* raises the activity coefficient of the monomer, and this causes the observed decrease in the equilibrium concentration of the monomer.

The effect of polymer concentration on $[M]_e$ has also been shown in the study of TOBOLSKY et al. (*35*). Unfortunately, their experimental technique was not sufficiently refined, and the wide scatter of experimental points prevented quantitative deductions from their data. Comment is necessary with respect to one statement made in their paper. The authors assumed that the "true" value of $[M]_e$ could be derived by linear extrapolation from the experimental $[M]_e$'s, obtained at high polymer concentrations, up to the point of intersection with the line $[M]_e = [M]_0$. However, this extrapolation is not valid because, as has been shown by VRANCKEN et al. (*13*), the results depend on the concentration of living ends. The proper determination of the "true" $[M]_e$ requires studies at low concentrations of living ends, as was done by WORSFOLD and BYWATER (*12*) and by McCORMICK (*11*).

In a recent paper, IVIN and LEONARD (*36*) considered the effect of a soluble polymer in a liquid monomer on the free energy of polymerization. The process may be represented by three steps: (1) remove a mole of monomer from the solution $(-\Delta \overline{G}_M)$, (2) convert the monomer into polymer $(\Delta G_{1,\,s})$, and (3) dissolve the polymer in the solution $(\Delta \overline{G}_P)$. In view of the equilibrium established between the monomer and polymer

$$-\Delta \overline{G}_M + \Delta G_{1,\,s} + \Delta \overline{G}_P = 0\,.$$

Accepting the conventional Flory-Huggins treatment, one deduces then the free energy of polymerization, $\Delta G_{1,s}$, of a liquid monomer into solid monomer to be

$$\Delta G_{1,s} = RT \{\ln \phi_1 - (\ln \phi_2)/n + 1 - 1/n + \chi(\phi_2 - \phi_1)\}.$$

For a high molecular weight polymer, i.e. for a large n, this equation is reduced to

$$\Delta G_{1,s} = RT \{\ln \phi_1 + 1 + \chi(\phi_2 - \phi_1)\}.$$

This equation is applicable to such a system as liquid tetrahydrofuran-dissolved polytetrahydrofuran which was discussed in part 3 of this review.

The thermodynamic relations are particularly simple for a polymer which is insoluble in an investigated solution of the monomer. For such a system

$$\Delta G_{1,m} + RT \ln a_m = \Delta G_{s,p}$$

for a living polymer which is in equilibrium with its monomer. The subscripts $1, m$ and s, p refer to liquid monomer and solid polymer, respectively.

An interesting example of such a system is a mixture of isobutene-sulfur dioxide and of the respective polysulfone (the alternating co-polymer of these two monomers) which is insoluble in the liquid monomers. For each composition of such a mixture there is a ceiling temperature, T_c, at which the system is at equilibrium. One expects a monotonic relation between $RT \ln\{a_{ib} \cdot a_{SO_2}\}$ and the respective T_c, because different values of the activities product should correspond to different ceiling temperatures. However, studies of Cook, Ivin, and O'Donnell (37a) led to a most surprising result. Using the activity coefficients determined in a previous work (37b), they constructed a curve, shown in Fig. 14, giving $\log\{a_{ib} \cdot a_{SO_2}\}$ as a function of $1/T_c$. This plot reveals that the respective function is not singled value, proving that different polymers are formed at different compositions of the mixture. The copolymers may differ in their composition. However, the elementary analysis showed that this is not the case; all the polymers had composition corresponding to 1 : 1 ratio of the monomers. Alternatively, they may differ in their structure or conformation. Both changes could be reflected in the crystal structure if the polymer is crystalline.

The authors prefered the latter explanation. They argue that the difference in the dielectric constant of the medium, when a hydrocarbon rich mixture is compared with that rich in sulfur dioxide, affects the ratio of gauche to trans conformations of the polymer chain. Such effects were observed in other systems, e.g. in 1,2 dichloroethane (38). Hence, the minute amount of living polymer, which is still in solution, should be intrinsically different in both media. However, the precipitated polymer may be ultimately the same, from whatever mixture it had been deposited, if the relaxation time is not too long. Since the glass transition of the polysulfone is rather high, it is possible that the ultimate equilibrium structure of the solid has not been attained, and therefore

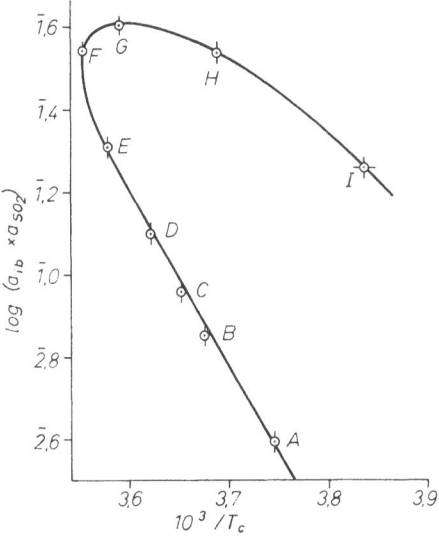

Fig. 14. Temperature dependence of the product of activities of isobutene, a_{ib} and SO_2, aSO_2, in equilibrium with its respective co-polymer. Reproduced, with permission, from Cook, Ivin and O'Donnel: Trans. Faraday Soc. **61**, 1887 (1965).

samples precipitated from solutions having different composition would exhibit a different x-ray pattern. This indeed has been observed (39).

In conclusion, studies of the equilibria between living polymers and their monomers provide means of determining monomer-polymer interaction parameters, and interaction parameters of monomer and polymer with a solvent.

11. Effect of pressure on the equilibrium constant of propagation

Polymerization leads to a contraction in the volume of the system so that the equilibrium of a monomer-polymer system shifts in the direction of the reaction as the hydrostatic pressure increases. This was demonstrated by Weale (40) in his studies of α-methylstyrene polymerization under high pressure. The ceiling temperature increased from 61° C at 1 atm to 170° C at 6480 atm.

A polymerization which cannot be observed under normal pressure might become suitable for study at elevated pressures. Consider, e.g., the case of a polymerization prohibitively slow at low temperatures and thermodynamically impossible at higher temperatures. Nonetheless, because the ceiling temperature should rise at higher pressure, the polymerization may become feasible under these conditions.

Consider now a polymerization which is thermodynamically impossible at normal pressure because the reaction proceeds with a positive ΔH and a negative ΔS. However, at higher pressures, ΔH may become negative and the process would then be feasible. This and similar problems have been reviewed in a recent paper by Ivin (41), [see also the review by Weale (42)].

Equilibrium between living polymer and its monomer offers excellent opportunities for studies of systems which cannot be polymerized under normal pressure. For example, steric strain prevents the polymerization of 1,1-diphenylethylene at atmospheric pressure, but the reaction might take place at a sufficiently high pressure. It should be possible to study such an equilibrium in a following device. Two optical cells, introduced into a chamber maintained at high pressure, are inserted into the optical paths of a two-beam spectrophotometer. One contains a solution of 1,1-diphenylethylene, the other has an identical solution to which some living dimer of 1,1-diphenylethylene has been added. The change in the optical density of the monomer could then be examined at different temperatures and pressures and this provides information about the thermodynamics of the polymerization reaction. This description is given to focus the reader's attention on the potentialities of such studies.

12. Equilibrium copolymerization

Copolymerization of monomers A and B involves steps

$$-\!\!-A + B \underset{k_{-12}}{\overset{k_{12}}{\rightleftarrows}} -\!\!-A B$$

and

$$-\!\!-B + A \underset{k_{-21}}{\overset{k_{21}}{\rightleftarrows}} -\!\!-B A$$

in addition to the homopropagation of A and B respectively. For some systems it was possible to determine directly the absolute rate constants of copolymerization (43, 44, 45) and in two cases even the rate constant of depropagation (46, 47). The addition of α-methylstyrene to living poly-2-vinylnaphthalene was shown to proceed reversibly (46) and under conditions of those experiments, the polymerization of further units of α-methylstyrene were virtually prevented. The reaction was studied in a stirred-flow reactor using radioactive monomer to simplify the analysis of the residual monomer. The method, developed in studies of the reac-

tion α-methylstyrene dimer $+$ monomer $\rightleftarrows$ trimer (23), (see p. 475), led
to the evaluation of k_{12}, k_{-12} and K_{12}.

The addition of styrene (S) to a polymer possessing $CH_2 . C(Ph)_2^-$, Na^+
end group was also investigated in a stirred-flow reactor (47). The process
involves two steps

$$—C(Ph)_2^-, Na^+ + S \underset{k_{-21}}{\overset{k_{21}}{\rightleftarrows}} —C(Ph)_2 . S^-, Na^+, K_{21}$$

$$—C(Ph)_2 . S^-, Na^+ + S \overset{k_{11}}{\rightarrow} —C(Ph)_2 . S . S^-, Na^+, \text{etc.}$$

The balance equations led to the relation

$$\{(S_0 - S_t)/S_t^3 \cdot D_t^-\} \{1 + (k_{11}/k_{-21}) S_t\} = K_{21} k_{11}^2 t^2 ,$$

where t denotes the residence time, S_0 the initial concentration of styrene,
and S_t and D_t^- the concentration of styrene and of $—C(Ph)_2^-$, Na^+,
respectively, in the stirred reactor. From this relation the value of K_{21}
was found to be about $5 \cdot 10^{-2}$ l/mole.

In a system containing living polymers, the propagation and depropagation continue until an equilibrium is established with respect
to *all* components. Therefore, for a living copolymer of A and B, the
ultimate distribution of the comonomers in the chain is given by the
equilibrium determined by the partition functions of the AA, BB, and
AB linkages. Calculation of the ultimate equilibrium distribution has
been recently reported by ALFREY and TOBOLSKY (48) who pointed out
that for an infinitely long polymer, the mathematical formalism of the
problem is identical with that developed by ISING (49) in his treatment
of ferromagnetism.

Denoting by N_A and N_B the number of A and B units and by M_{AB}
the number of AB linkages, one finds that

$$(N_A - M_{AB}) (N_B - M_{AB})/(M_{AB})^2 = (f_{AA} \cdot f_{BB}/f_{AB}^2) \exp. (\Delta E_{AB}/RT)$$

f_{AA}, f_{BB}, and f_{AB} denote the partition functions of the respective linkages
and $\Delta E_{AB} = 2 E_{AB} - E_{AA} - E_{BB}$ where E's are the energies of the
respective bonds.

The distribution of sequence lengths is given by the equations

$$n_x^A = N_A p_{AA}^{x-1} (1 - p_{AA})^2$$
$$n_x^B = N_B p_{BB}^{x-1} (1 - p_{BB})^2$$

where n_x^A and n_x^B are the numbers of A or B segments having length x,
and p_{AA} and p_{BB} are defined by

$$p_{AA} = (N_A - M_{AB})/N_A \text{ and } p_{BB} = (N_B - M_{AB})/N_B .$$

It should be stressed that the initial copolymer formed by the addition
of two monomers to an ionic initiator is expected to be a block polymer
because in ionic polymerization it is common to find k_{AA} and k_{BA}, i.e.

the rate constants of A addition to units terminated by A or B, to be much greater than either k_{BB} or k_{AB}. This rule is not perfectly general and exceptions are known. The redistribution of monomers to form the ultimate equilibrium copolymer would be usually a slow process, see, e.g., section 8, p. 480, and probably it will not be achieved in any practical experiment.

Acknowledgment. Studies of the Syracuse group were supported by grants from the National Science Foundation.

The writer is indebted to Professor F. S. Dainton, Dr. K. J. Ivin, and to Dr. H. A. Skinner for reading the manuscript and for their constructive comments.

References

1. Dainton, F. S., and K. J. Ivin: Nature **162**, 705 (1948).
2. — — Quart. Rev. (London) **12**, 61 (1958).
3. — — Discussions Faraday Soc. **14**, 199 (1953).
4. Cook, R. E., F. S. Dainton, and K. J. Ivin: J. Polymer Sci. **26**, 351(1957).
5. Dainton, F. S., J. Dieper, K. J. Ivin, and D. R. Sheard: Trans. Faraday Soc. **53**, 1269 (1957).
6. Ivin, K. J.: Trans. Faraday Soc. **51**, 1273 (1955).
7. Bywater, S.: Trans. Faraday Soc. **51**, 1267 (1955).
8. Tobolsky, A. V.: J. Polymer Sci. **25**, 220 (1957); **31**, 126 (1958).
9. —, and A. Eisenberg: J. Am. Chem. Soc. **82**, 289 (1960).
10. Szwarc, M.: Proc. Roy. Soc. (London); Ser. A, **279**, 260 (1964).
11. McCormick, H. W.: J. Polymer Sci. **25**, 488 (1957).
12. Worsfold, D. J., and S. Bywater: J. Polymer Sci. **26**, 299 (1957).
13. Vrancken, A., J. Smid, and M. Szwarc: Trans. Faraday Soc. **58**, 2036 (1962).
14. Bywater, S., and D. J. Worsfold: J. Polymer Sci. **58**, 571 (1962).
15. Meerwein, H., D. Delfs, and H. Morschel: Angew. Chem. **72**, 927 (1960).
16. Bawn, C. E. H., R. M. Bell, and A. Ledwith: Polymer (London) **6**, 95 (1965). Communicated at the Anniversary Meeting of the Chemical Society, London, in Cardiff, 1963.
17. Vofsi, D., and A. V. Tobolsky: J. Polymer Sci. **3** A, 3261 (1965).
18a. Dreyfuss, M. P., and P. Dreyfuss: Polymer (London) **6**, 93 (1965).
18b. — — J. Polymer Sci. **4**A, 2179 (1966).
18c. Sims, D. J.: J. Chem. Soc. (London) p. 864 (1964).
19. Frank, C. F.: J. Org. Chem. **26**, 307 (1961).
20. Lee, C. L., J. Smid, and M. Szwarc: J. Phys. Chem. **66**, 904 (1962).
21. Bergmann, E., H. Taubadel, and H. Weiss: Chem. Ber. **64**, 1493 (1931).
22. Vrancken, A., J. Smid, and M. Szwarc: J. Am. Chem. Soc. **83**, 2772 (1961).
23. Lee, C. L., J. Smid, and M. Szwarc: J. Am. Chem. Soc. **85**, 912 (1963).
24a. Denbigh, K. J.: Trans. Faraday Soc. **40**, 352 (1944).
24b. Stead, B. F. M. Page, and K. J. Denbigh: Discussions Faraday Soc. **2**, 263 (1947).
25. Brown, W. B., and M. Szwarc: Trans. Faraday Soc. **54**, 416 (1958).
26. Flory, P. J.: Principles of Polymer Chemistry, Cornell University Press (1953); a) p. 318; b) p. 336.
27. Hermans, J. J.: J. Polymer Sci. **4** C, 345 (1966).
28. Tobolsky, A. V., and A. Eisenberg: J. Colloid. Sci. **17**, 49 (1962).
29. Dainton, F. S., and R. H. Tomlinson: J. Chem. Soc. (London) 151 (1953).

30. GEE, G.: Trans. Faraday Soc. 48, 515 (1952).
31. TOBOLSKY, A. V., and A. EISENBERG: J. Am. Chem. Soc. 81, 780 (1959).
32. EISENBERG, A.: J. Polymer Sci. B 1, 33 (1963).
33. SCOTT, R. L.: J. Chem. Phys. 17, 268 (1949).
34. BYWATER, S.: Makromol. Chem. 52, 120 (1962).
35. TOBOLSKY, A. V., A. REMBAUM, and A. EISENBERG: J. Polymer Sci. 45, 347 (1960).
36. IVIN, K. J., and J. LEONARD: Polymer (London) 6, 621 (1965).
37a.COOK, R. E., K. J. IVIN, and J. H. O'DONNELL: Trans. Faraday Soc. 61, 1887 (1965).
37b.AYSCOUGH, P. B., K. J. IVIN, and J. H. O'DONNELL: Trans. Faraday Soc. 61, 1601 (1965).
38. WADA: J. Chem. Physics 22, 198 (1954).
39. IVIN, K. J., and J. H. O'DONNELL: Trans. Faraday Soc. (in press).
40. KILROE, J. G., and K. E. WEALE: J. Chem. Soc. (London) 3849 (1960).
41. IVIN, K. J.: Pure and Applied Chem. 4, 271 (1962).
42. WEALE, K. E.: Quart. Rev. (London) 16, 267 (1962).
43. BHATTACHARYYA, D. N., C. L. LEE, J. SMID, and M. SZWARC: J. Am. Chem. Soc. 85, 533 (1963).
44. SHIMA, M., D. N. BHATTACHARYYA, J. SMID, and M. SZWARC: J. Am. Chem. Soc. 85, 1306 (1963).
45. —, J. SMID, and M. SZWARC: J. Polymer Sci. 2 B, 735 (1964).
46. STEARNE, J., J. SMID, and M. SZWARC: Trans. Faraday Soc. 60, 2054(1964).
47. URETA, E., J. SMID, and M. SZWARC: J. Polymer Sci. 44, 2219 (1966).
48. ALFREY, T., and A. V. TOBOLSKY: J. Polymer Sci. 38, 269 (1959).
49. ISING, E.: Z. Physik 31, 253 (1925).
50. POULIS, J. A., C. H. MASSEN, and D. V. D. LEEDEN: Trans. Faraday Soc. 58, 474 (1962).
51. —, and W. DERBYSHIRE: Trans. Faraday Soc. 59, 559 (1963).
52. —, C. H. MASSEN, A. EISENBERG, and A. V. TOBOLSKY: J. Am. Chem. Soc. 87, 413 (1965).

Received February 14, 1966

Adv. Polymer Sci., Vol. 4, pp. 496—527 (1967)

Aromatic Polyethers

By

A. S. HAY

General Electric Research and Development Center
Schenectady, New York

With 4 Figures

Table of Contents

I. Introduction

Low molecular weight aromatic ethers have been prepared principally by the condensation of phenolate salts with aromatic halides (*82*). The Ullmann condensation (*87*), which employs copper or its salts as catalysts has been used in most cases in the laboratory. Recently a modification of the Ullmann condensation which consists of heating copper (1) oxide, the free phenol, and the aromatic halide in s-collidine has been reported (*3*). This method is recommended for alkali-sensitive aromatic compounds. In addition, reaction of phenolate salts with copper (1) oxide and the aromatic halide in boiling N,N-dimethyl formamide is described. When the halogen is activated by electronegative groups as in *p*-chloronitrobenzene or 2.4-dinitrochlorobenzene, no catalyst is necessary, however aprotic, dipolar solvents are now commonly used to speed up reactions of this type.

Diphenyl ether itself is a byproduct in the manufacture of phenol from chlorobenzene and aqueous caustic at elevated temperatures (29, 30, 31, 32, 33) and by proper control of conditions can be made the major product of the reaction.

Another method for the preparation of aromatic ethers which has received relatively little attention because of experimental difficulties and because it is not a general synthesis, was pioneered by SABATIER (68, 69, 70, 71). This involved the vapor phase dehydration of phenols over alumina or preferably thoria at temperatures in the neighborhood of 400° C.

$$2 \left\langle \overline{\bigcirc} \right\rangle\text{-OH} \xrightarrow[\triangle]{\text{ThO}_2} \left\langle \overline{\bigcirc} \right\rangle\text{-O-}\left\langle \overline{\bigcirc} \right\rangle + \text{H}_2\text{O}$$

BRINER has studied this reaction further and showed that it was reversible (7, 8). The reaction proceeds in good yield for simple phenols without substituents in the o-positions and appears to be the basis also for a potential commercial synthesis of diphenyl ether (13). Low yields of aromatic ethers have also been obtained by pyrolysis of a variety of phenolate salts and esters (27, 41).

Naphthols are readily dehydrated to the corresponding ethers at considerably lower temperatures (300° C) in the liquid phase in the presence of metal oxides (16, 17).

In both of the foregoing cases when the reactions are performed at higher temperatures, dibenzofurans or dinaphthofurans become the major products.

An extensive review of the preparation and properties of aromatic ethers is available (82).

Aromatic ethers are notable for their exceptional thermal, oxidative, and hydrolytic stability. Because of this stability they have been synthesized for use as high temperature fluids (28, 72).

High molecular weight, linear, aromatic ether polymers have recently been prepared by a variety of methods. This review will cover the principal methods that have been used to prepare this new class of high molecular weight polymers. The chemistry of the polymerization reactions involved as well as the properties of the resulting polymers will be described.

II. Synthesis of aromatic polyethers

High molecular weight aromatic ether polymers have now been successfully prepared by a number of different methods and several other methods have been unsuccessfully explored. In order to more completely describe the individual synthetic methods the experimental details of a representative example are included where possible in the cases where high polymers have been successfully prepared.

1. Nucleophilic substitution of halogen

A. Ullmann condensation

This is perhaps the most ovious approach to the synthesis of a high molecular weight aromatic ether because it is the reaction which has most commonly been used in the laboratory. However, the yields that have been reported for the preparation of simple aromatic ethers by this method do not appear to be high enough to allow the formation of high molecular weight polymers (72, 82).

GOLDEN obtained polymeric materials with molecular weights up to 1800 from potassium p-chlorophenoxide in refluxing nitrobenzene as solvent in the presence of copper bronze as catalyst (28).

BROWN has reported that poly-1.3-phenylene ether (I) with a molecular weight in the neighborhood of 10,000 can be prepared by heating alkali metal salts of m-bromophenol, in solution, in the presence of copper metal or copper salts. Generally, the reactions were performed between 110° C and 200° C in solvents such as pyridine, nitrobenzene and phenyl ether (9, 10).

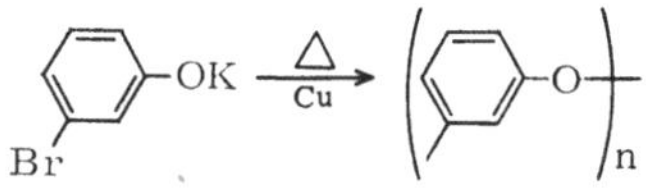

I

Similar results were obtained with p-bromophenol (11). In some of the preparations described, cyclics were minor products of the reaction.

Recently, in a United States patent (75), STAMATOFF has claimed the preparation of high molecular weight poly-1.4-phenylene ether by a modification of the Ullmann condensation applied to sodium p-bromophenoxide. The catalyst used in this case was usually copper (I) chloride in conjunction with a variety of organic bases (pyridine; N, N-dimethylformamide, N,N-dimethylacetamide or benzothiazole). By rigorous exclusion of oxygen and moisture at elevated temperatures (200° C) in solvents such as dimethoxybenzenes, nitrobenzene or benzophenone, it is claimed that poly-1.4-phenylene ether is obtained. No data on molecular weight or intrinsic viscosity or other physical properties of the resulting polymers is given in the patent. It is claimed that the polymers can be melt spun into fibers or molded in the temperature range 300—350° C thus implying that the polymers are high molecular weight.

Experimental

a) Preparation of sodium p-bromophenolate

Into a 1 liter flask was placed 500 ml methanol. This was sparged with nitrogen to remove the air from the flask. Sodium (17.38 g, 0.756 mole) was then added, and the mixture was stirred and cooled by an ice bath. After the sodium had dissolved, p-bromophenol (130.80 g, 0.756 mole) was added. The mixture was heated by means of a boiling water bath while passing nitrogen through the methanol, until the solid sodium salt of p-bromophenol remained in the flask. This was further dried by heating at 100° C under reduced pressure.

b) Preparation of the catalyst

Into a bottle in an atmosphere of nitrogen was placed 0.2 g of cuprous chloride and several clean pieces of copper were added. To this was added 20 ml of pyridine. The mixture was agitated, forming a light brownish solution. The catalyst solution was very sensitive to oxygen. The solution was evaporated to dryness, leaving a complex of cuprous chloride with pyridine.

c) Polymerization

Into a glass tube having a diameter of 1 inch and 13 inches in length, a stream of dry helium was directed. Into the oxygen free tube was injected 10 g of p-diethoxybenzene and 10 g of sodium p-bromophenolate. The contents of the tube were freed of any further quantity of oxygen by evacuating the tube three times and replacing the atmosphere with helium. The tube was then heated to a temperature of 168° C and a catalyst consisting of .01 g of cuprous chloride in 1 ml of pyridine was injected. Heating was continued for 20 hours. The contents of the tube were cooled and removed from the tube by means of acetone in large excess. The resulting mixture was treated with further quantities of acetone and methanol in a Waring Blendor and washed several times with boiling dilute hydrochloric acid. The resulting polymer was washed again with methanol and water and finally with a methanol-acetone mixture. The resulting polymer when dried in a drying oven was obtained in the form of a white powder.

B. Nucleophilic substitution of activated halogens

High molecular weight polymers have been recently prepared by condensation of alkali metal salts of biphenols with activated halides such as 4.4'-dichlorodiphenyl sulfone (50). The polymer (II) prepared from 4.4'-isopropylidenediphenol and 4.4'-dichlorodiphenylsulfone is now commercially available.

II

Table 1

HO–R–OH	X–R'–X	Tg(°C)
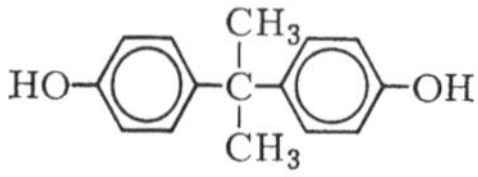 HO–C6H4–C(CH3)2–C6H4–OH	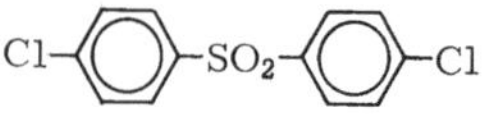 Cl–C6H4–SO2–C6H4–Cl	190
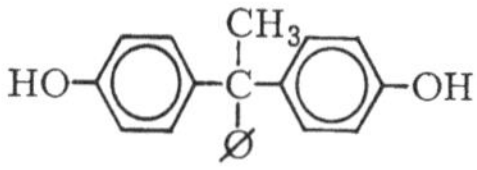 HO–C6H4–C(CH3)(Ø)–C6H4–OH		200
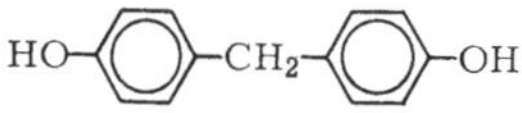 HO–C6H4–CH2–C6H4–OH		180
 HO–C6H4–OH		—[1]
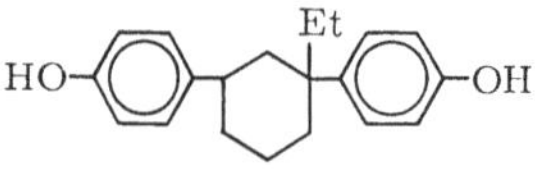 HO–C6H4–C(cyclohexane)(Et)–C6H4–OH	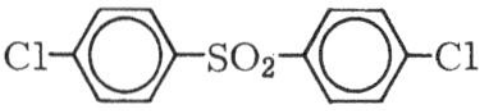 Cl–C6H4–SO2–C6H4–Cl	230
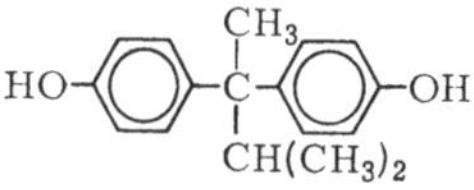 HO–C6H4–C(CH3)(CH(CH3)2)–C6H4–OH		200
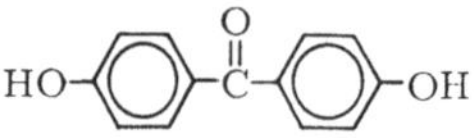 HO–C6H4–C(=O)–C6H4–OH		205
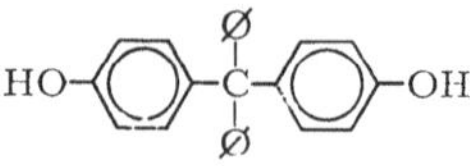 HO–C6H4–C(Ø)(Ø)–C6H4–OH		230
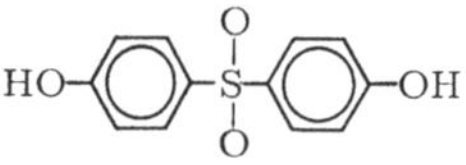 HO–C6H4–S(=O)(=O)–C6H4–OH		—
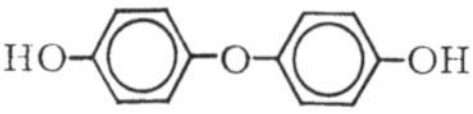 HO–C6H4–O–C6H4–OH		180
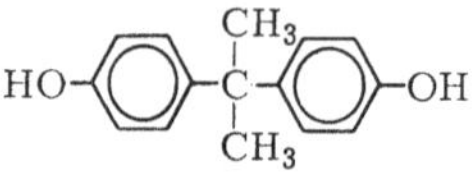 HO–C6H4–C(CH3)2–C6H4–OH	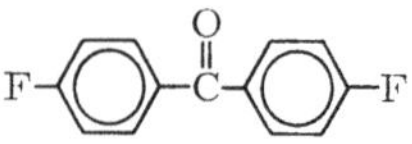 F–C6H4–C(=O)–C6H4–F	155
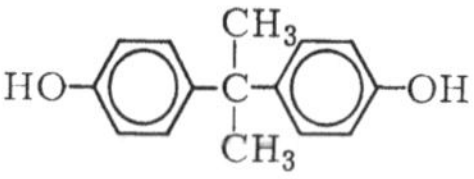 HO–C6H4–C(CH3)2–C6H4–OH	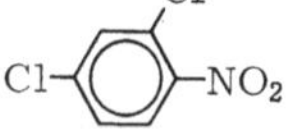 Cl–C6H3(Cl)–NO2	150
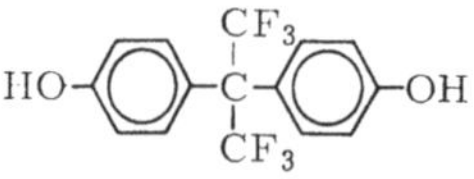 HO–C6H4–C(CF3)2–C6H4–OH	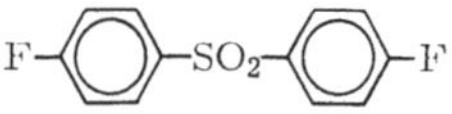 F–C6H4–SO2–C6H4–F	205
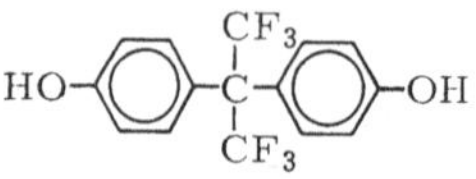 HO–C6H4–C(CF3)2–C6H4–OH	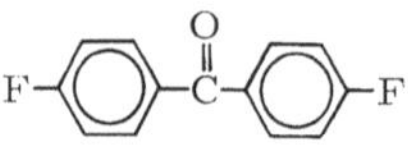 F–C6H4–C(=O)–C6H4–F	175

---

[1] Softening Temp. = 310° C.

The reaction is carried out in aprotic dipolar solvents such as N,N-dimethyl formamide, dimethyl sulfone or dimethyl sulfoxide. Oxygen and water are rigorously excluded from the reaction which proceeds to completion in the neighborhood of 130—140° C in four to five hours. The reaction has been applied to a large number of other biphenols (Table 1) and in addition activated dihalides such as 2.4-dichloronitrobenzene and 4.4-di-fluorobenzophenone have been used successfully.

Experimental

To a 250 ml flask equipped with a stirrer, a thermometer, a water condenser and Dean-Stark trap was added 11.42 g of 4.4'-isopropylidenediphenol (0.05 moles), 13.1 g of a 42.8% potassium hydroxide solution (0.1 mole KOH), 50 ml of dimethyl sulfoxide and 6 ml of benzene. The reaction mixture was kept under an atmosphere of nitrogen and the water was azeotroped off over a 3 to 4 hours period (130—135°C). At the end of this time the reaction mixture consisted of the potassium salt of the biphenol and was essentially anhydrous. After cooling the mixture there was added 14.35 g (0.05 moles) of 4.4'-dichlorodiphenyl sulfone and 40 ml of anhydrous dimethylsulfoxide. The reaction mixture was maintained, under a nitrogen atmosphere, between 130 and 140° C with stirring for 4 to 5 hours. The viscous orange solution was then poured into 300 ml of water in a Waring Blendor and the polymer separated by filtration and dried at 110° for 16 hours. A yield of 22.2 g (100%) of polymer with a reduced viscosity in chloroform (0.2 g per 100 ml at 25°) of 0.59 was obtained.

2. Oxidative displacement of halogen

Historically the first one to have prepared an aromatic ether of any appreciable molecular weight was HUNTER.

In 1911, HUNTER described the preparation and properties of silver salts of some halogenated phenols (80). He found that the silver salt of 2.4.6-tribromophenol reacted with ethyl iodide to give a deep blue solution which gradually faded to a brownish yellow. There remained a precipitate of silver bromide and a solution which when added to alcohol precipitated an amorphous product.

Subsequently, in a series of papers, HUNTER and coworkers investigated the reaction more thoroughly and attempted to characterize this amorphous product (43, 44, 45, 46, 47, 48, 88). He reported that this material had a molecular weight of 6,600 (duplicate determination, 12,400) and analyzed for $C_6H_2Br_2O$. Hence he concluded the product was a polymer and he described the polymerization reaction as follows:

$$n \; C_6H_2Br_3OAg = (C_6H_2Br_2O)n + n \; AgBr$$

Similar products were obtained in the decomposition of the silver salt of 2.4.6-trichlorophenol, however, in this case higher temperature (60° C) was necessary and ethyl iodide did not initiate the reaction. Polymers were also obtained from several other trihalophenols.

A similar polymer was obtained by treating 2.4.4.6-tetrabromo-1.4-cyclohexadienone (III) with mercury (48).

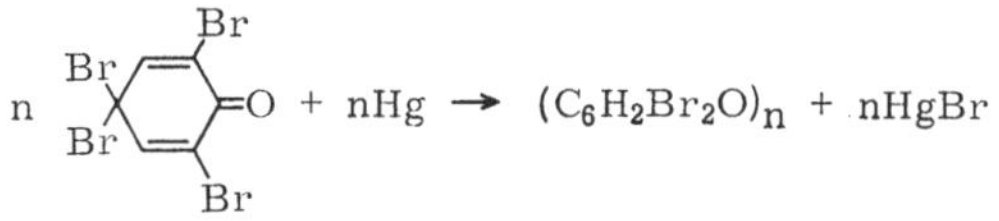

$$n \quad \overset{Br}{\underset{Br}{\overset{Br}{\diagdown}}}\!\!=\!\!O + nHg \rightarrow (C_6H_2Br_2O)_n + nHgBr$$

III

GOLDEN (28) has prepared a large number of polymers of this type by modifications of Hunter's method (Table 2). All of the polymers he obtained yielded brittle films when cast from solution as would be expected for low molecular weight, branched polymers. GOLDEN also noted the possibility of preparing dibenzodioxanes (IV) and in the decomposition of sodium pentachlorophenoxide he was able to control the reaction so that either product could be obtained.

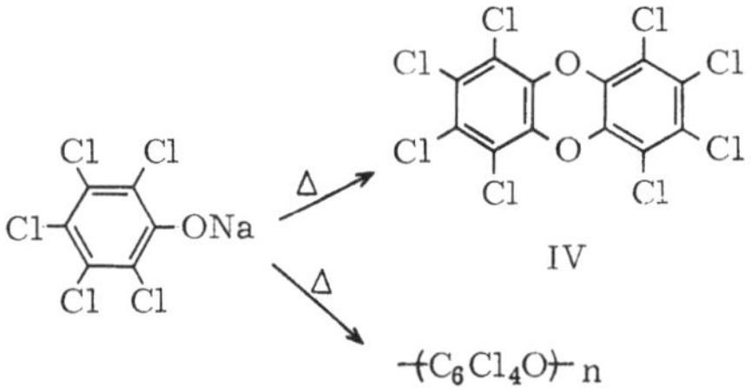

IV

$\cdot(C_6Cl_4O)\cdot n$

V

Table 2

Parent phenol	Phenoxide employed	Polymer		
		Yield (%)	Softening pt. (°C)	Mol. Wt.
4-Chloro	K	4	256—260	2800
2.4-Dichloro	Na	45	270—273	—
2.4-Dibromo	Na	81	240—245	—
2.4.5-Trichloro	Na	37	215—220	—
2.4.6-Trichloro	Na	90	185—188	4800
	Ag	79	185—188	9600
2.4.6-Tribromo	K	60	244—246	3200
2.3.4.6-Tetrachloro	Na	78	219—222	—
	Ag	80	230—235	—
Pentachloro	Ag	73	192—194	1500
Pentabromo	Na	63	300—310	—

The polymerization of 2.4.6-trichlorophenol with lead oxide and silver oxide has also been investigated by HEDAYATULLAH and DENIVELLE (40) and MÜLLER (65) has also oxidized a number of polyhalophenols and looked at the intermediates in the reaction in order to develop a mechanism for the reaction.

PRICE and coworkers (14) found that the polymerization of 2.6-dimethyl-4-bromophenol could be initiated in the presence of base with a variety of oxidizing agents (e.g., $K_3Fe(CN)_6$, PbO_2). Much earlier, AUWERS (1, 2), during a study of the preparation of 3.3'.5.5'-tetraalkyl-diphenoquinones by oxidation of 2.6-dialkylphenols noted that in the oxidation of 2.6-dimethyl-4-bromophenol with potassium ferricyanide only an amorphous product was obtained. They did not investigate the product further; however, it undoubtedly was a low molecular weight polyphenylene ether. PRICE found that about 10 mole percent of the initiator was necessary for high conversions. Initially they obtained polymers with molecular weights less than 10,000 (74) but later they successfully prepared high molecular weight polymers (67).

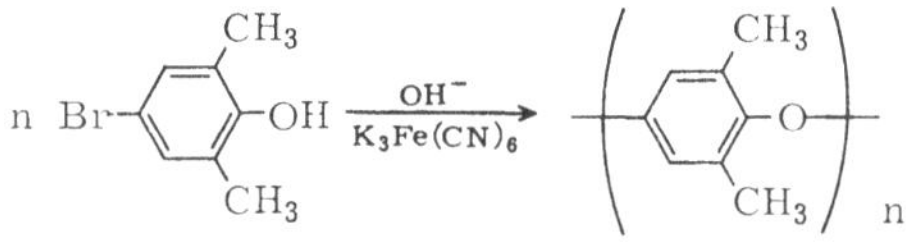

The reaction is extremely rapid, being substantially complete in less than one minute at room temperature. A similar reaction occurs with 2.6-dimethyl-4-chlorophenol. KURIAN and PRICE (53) also polymerized several other 2.6-disubstituted-4-halophenols (Table 3). Only low molecular weight polymers (as measured by intrinsic viscosity) appear to have been obtained.

Table 3

4-Bromophenol	Intrinsic viscosity[1]
2.6-Diallyl-	0.189
2.6-Dipropyl-	0.094
2-Propyl-6-phenyl-	0.076
2-Allyl-6-phenyl-	0.108
2-Allyl-6-methyl-	0.168
2-Propyl-6-methyl-	0.168

[1] Highest of several runs reported.

However, high molecular weight polymers might be attainable from most of these monomers by optimization of the individual reaction.

Experimental (13)

A 1 l. three-necked flask is fitted with an efficient stirrer, a dropping funnel, and a gas inlet tube connected to a stream of purified nitrogen. A solution of 5 g of potassium hydroxide in 200 ml of water, 8 g (0.04 mole) of 4-bromo-2.6-xylenol and 200 ml of benzene is introduced. The stirrer is started and 1.3 g of potassium ferricyanide in 20 ml of water is added dropwise over a period of 30 minutes. After an additional 15 min of stirring, the mixture is transferred to a separatory funnel and the aqueous phase is drawn off the bottom. The yellow benzene solution is transferred to a 300 ml distilling flask and concentrated to 50 ml under water pump pressure. The concentrate is poured slowly, with stirring, into 250 ml of methanol acidified with 2.5 ml of concentrated hydrochloridc acid. The precipitate is collected by suction filtration and washed by resuspending in 150 ml of methanol. It is then collected, redissolved in 50 ml of benzene, and precipitated again as described above.

The reprecipitated polymer is collected and redissolved in 50 ml of benzene contained in a 250 ml round-bottomed flask. The flask is swirled in Dry-Ice acetone until the contents are frozen and it is quickly connected to a vacuum line protected by a large trap. The pump is started and a pressure of 1 mm Hg is maintained for 5 hours. Yield: 4.6 g (96%) of spongy, white solid, $[\eta]_{C_6H_6}^{30\,°C} = 0.5-0.6$ dl/g. The product darkens and softens above 270° C.

In a recent patent STAMATOFF (76) has described the polymerization of unsymmetrically substituted halophenols in a manner similar to that described by PRICE, using as initiators peroxides or persulfates.

More recently, (26) salts of para-bromophenols have been reacted with a variety of oxidizing agents as initiators under anhydrous conditions in the presence of aprotic, dipolar solvents. Under these conditions it is remarkable that the product from 2,6-dichloro-4-bromophenol appears to be the high molecular weight, linear 1.4-phenylene ether (VI).

VI

Experimental

To a solution of 40 g (1.0 moles) of sodium hydroxide in 500 ml of methanol was added 242 g (1.0 moles) of 2.6-dichloro-4-bromophenol. The pH was adjusted between 9.0 and 10.0 (preferably 9.5) by means of one or another of the reactants. The pH was determined by diluting a 2.5 g aliquot with 100 ml of 50% aqueous methanol. The alcohol and water were removed by distillation. In a one liter round bottom flask there was introduced 100 g of the sodium salt of 2.6-dichloro-4-bromophenol, 350 ml of chlorobenzene and 40 ml of N,N-dimethylformamide. The mixture was agitated until the salt was in solution then immediately there was added 26 ml of dimethylsulfoxide. A suspension forms. The air was removed by alternate evacuation and introduction of nitrogen then there was added 1.0 g of benzoyl peroxide dissolved in 10 ml of toluene. The mixture was stirred for 80 min at 29−33° C then for 5 hours at 54−59° C. The formation of polymer was indicated by the disappearance of the particles of the suspension and an increase in the viscosity of the solution. The polymer was isolated by precipitation into acetone. After filtration the polymer was washed thoroughly with water, then with acetone and then dried at 100° C. There was obtained 60 g (theoretical) of poly-(2.6-dichloro-1.4-phenylene ether).

Anal. Calcd. for C: 44.7; H, 1.3; Cl, 44.1; Br, 0.

Found: C, 44.6; H, 1.5; Cl, 41.9; Br, 1.1. The product can be compression molded at 350° C into a tough, almost colorless film.

3. Oxidative coupling

In 1959 HAY, et al., (36) reported that certain 2.6-disubstituted phenols reacted with oxygen in the presence of an amine complex of a copper salt as catalyst to yield high molecular weight polyphenylene ethers.

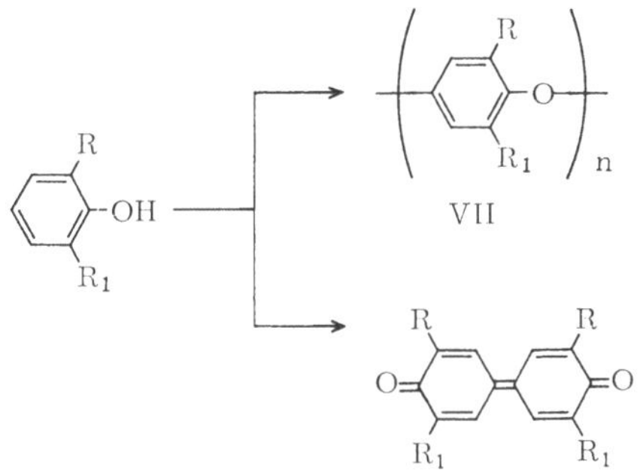

Diphenoqinones were concomitant products. This was the first synthesis of a linear polyphenylene ether with high enough molecular weight to have useful physical properties.

WATERS (39) has described the oxidation of 2.6-dimethylphenol with alkaline ferricyanide. The products he obtained were the diphenoquinone (VIII; $R=R_1=CH_3$) and an amorphous material (M.W. $\sim$ 800) which he did not further characterize. In retrospect, it would appear that this product was a low molecular weight polyphenylene ether (VII; $R=R_1=CH_3$).

Oxidation of 2.6-dimethylphenol with silver oxide in benzene solution has been shown by LINDGREN (58), to also yield a low molecular weight ($\sim$ 2000) polyether in low yield as well as the diphenoquinone.

Using activated manganese dioxide, silver oxide, or lead dioxide as the oxidizing agents, McNELIS has also obtained low molecular weight polymer from 2.6-dimethylphenol (60, 61).

KWIATEK has prepared (IX) (54)

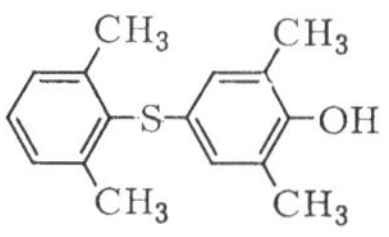

IX

and under the conditions of oxidative polymerization as described by HAY, the polyether (VII; $R=R_1=CH_3$) and the dixylyldisulfide are obtained. Similar results were obtained when (X) was oxidized (55).

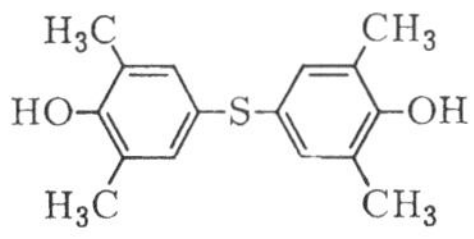

X

In this case the polyether was obtained and presumably sulfur was the other product.

The oxidative polymerization reaction is rapid at room temperature. Oxidation of 2.6-dimethylphenol readily gives high polymer with only a minor amount of the diphenoquinone (VIII; $R=R_1=CH_3$). This polymer is now being produced commercially. In general when the substituents are small (Table 4) the polymer is formed preferentially (35). If one of the substituents is as large as tert-butyl or both as large as isopropyl then the diphenoquinone is preferentially formed.

Table 4

R_1	R_2	polymer	diphenoquinone
CH_3	CH_3		
CH_3	CH_2CH_3	×	
CH_3	$CH(CH_3)_2$	×	
CH_3	$C(CH_3)_3$		×
CH_3	ϕ	×	
CH_3	OCH_3	×	
CH_3	Cl	×	
CH_2CH_3	CH_2CH_3	×	
$CH(CH_3)_2$	$CH(CH_3)_2$		×
$C(CH_3)_3$	$C(CH_3)_3$		×
Cl	Cl	×	
NO_2	NO_2	NR	
OCH_3	OCH_3		×

In the former case the diphenoquinone is formed exclusively while in the latter case small amounts of low molecular weight polymer have been observed. As would be expected, substituents which raise the oxidation potential of the phenol retard the polymerization. Thus whereas 2.6-dimethylphenol polymerizes readily at room temperature, temperatures in the neighborhood of 60° C are required to polymerize 2-chloro-6-methylphenol at comparable rates and even higher temperatures are necessary to oxidize 2.6-dichlorophenol.

Experimental (37)

To a 500 ml wide-mouthed Erlenmeyer flask in a water bath at 30° C equipped with a Vibromixer stirrer, an oxygen inlet tube, and a thermometer are added 200 ml of nitrobenzene, 70 ml of pyridine, and 1 g of copper (I) chloride. Oxygen (300 ml/min) is bubbled through the vigorously stirred solution and then 15 g (0.12 mole) of 2.6-dimethylphenol is added. Over a period of 16 min the temperature rises to 33° C, at which point the reaction mixture begins to get viscous. The reaction is continued for 12 min, then it is diluted with 100 ml of chloroform and added to 1.1 l of methanol containing 3 ml of concentrated hydrochloric acid. The precipitated polymer is filtered and washed with 250 ml of methanol, then with 250 ml of methanol containing 10 ml concentrated hydrochloric acid, and finally with

250 ml of methanol. The polymer is dissolved in 500 ml of chloroform, filtered and reprecipitated in 1.2 l of methanol containing 3 ml of concentrated hydrochloric acid. After washing with methanol and drying at 110° C (5 mm) for 3 hours there is obtained 13.5g (0.11 mole, 91%) of almost colorless polymer, $[\eta] = 0.96$ (dl/g in CHCl$_3$ at 25° C.

Polymerization also takes place when 4-halo-2.6-disubstituted phenols are oxidized with copper-amine catalysts and oxygen (**5,35**). In this case, stoichiometric amounts of copper salt or some other chloride acceptor (inorganic bases or strongly basic amines) are necessary since the amine complexes of copper (II) halides are not catalysts for the polymerization. BLANCHARD (**5**) has also described the polymerization of these 4-halo-phenols under conditions similar to those used by PRICE using certain copper (II) complexes as initiators.

High molecular weight polymers have also been prepared from a monosubstituted phenol, o-cresol, under special conditions (**38**), although in general the oxidation of phenols with open o-positions gives complex low molecular weight products (**66**). LINDGREN (**59**) has oxidized guaiacol with silver oxide and potassium ferricyanide and obtained amorphous materials with molecular weights below 1000.

4. Decomposition of benzene-1,4-diazooxides

The decomposition of benzene-1.4-diazooxide might be expected to give as an intermediate (XI) which could possibly polymerize to a poly-phenylene ether. In 1956, Süs studied the photochemical decomposition of benzene-1.4-diazooxide (*79*). He states

XI

that irradiation of thin films of the solid yields an insoluble material for which he proposes the phenylene ether structure. No experimental evidence was given to support this statement. In methanol as solvent, p-methoxyphenol was the product and p-arylphenols were obtained in aromatic solvents. However, WANG (*83*) has claimed that in benzene a high melting solid (300° C) is formed for which he too proposes the phenylene ether structure.

DEWAR (*21*) studied the thermal decomposition of 2.6-dibromo-benzene-1.4-diazooxide. It was hoped that by flanking the oxygen with bromine atoms that side reactions involving the positions o- to the oxygen might be minimized. Only low molecular weight polymers (1600 to 3600) were obtained. Analysis of the polymers showed a loss of bromine, hence it was concluded that in the propagation step some bromine atoms were displaced giving polymers of irregular structure.

Price has studied the thermal and photodecomposition of 3.5-dimethyl-1.4-diazooxide under a variety of conditions (52). No polymers were obtained except when the decomposition was run in cyclic ethers. Under these conditions low molecular weight materials in which considerable amounts of solvent were incorporated were obtained. Similar results have been obtained by Stille (78).

5. Friedel-Crafts condensation with sulphonyl chlorides

Very recently it has been reported that the Friedel-Crafts condensation of aromatic disulfonyl chlorides with aromatic ethers has been successfully effected to yield high molecular weight polymers (20)

$$\text{n } ClO_2SArSO_2Cl + \text{n } \langle\bigcirc\rangle\text{-O-}\langle\bigcirc\rangle \rightarrow \text{+}SO_2Ar\text{-}SO_2\text{-}\langle\bigcirc\rangle\text{-O-}\langle\bigcirc\rangle\text{+}_n$$

Stoichiometry is apparently better maintained by self-condensation of a sulfonyl chloride such as (XII)

$$\text{n } \langle\bigcirc\rangle\text{-O-}\langle\bigcirc\rangle\text{-}SO_2Cl \rightarrow \text{+}\langle\bigcirc\rangle\text{-O-}\langle\bigcirc\rangle\text{-}SO_2\text{+}_n$$

XII

Contrary to what might be expected the reaction requires only catalytic amounts of catalyst, preferably ferric chloride, and is carried out at elevated temperatures ($> 150°$ C) in the absence of solvent for extended periods of time.

It is interesting to note that by proper choice of monomers identical polymers could be obtained by this method and method IB previously discussed.

Experimental

Weighed quantities of anhydrous ferric chloride (purified by sublimation in a stream of chlorine) and the monosulphonyl chlorides were melted under a nitrogen atmosphere and stirred to dissolve the catalyst. The temperature was then raised to 190° C over 10 to 20 min, during which hydrogen chloride was evolved and the reaction mixture foamed. After cooling, the foamed mass was ground to a powder under nitrogen and then heated to ca. 200° C for 20 to 30 min when further evolution of hydrogen chloride occurred and the powder sintered. By the end of this stage over 75% of the theoretical quantity of hydrogen chloride had been evolved. The cooled polymer was then reground under nitrogen and heated to 250° C under a vacuum of 1.5 mm for 2 to 3 hours to complete the polymerization. After cooling, the product was treated with a one per cent solution of acetylacetone in dimethyl formamide at 100° C for 10 min, filtered to remove insoluble polymer, and the filtrate poured into acetone. The precipitated polymer was washed with acetone and then dried under vacuum at 120° C. Some of the polymers that have been prepared in this fashion are shown in Table 5.

Table 5

Sulphonyl chloride	FeCl$_3$	Final temp. °C	Yield	Insolubles* %	RV[1]
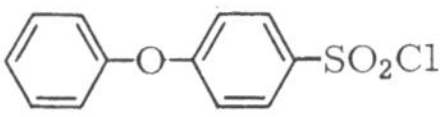 ⟨◯⟩-O-⟨◯⟩-SO$_2$Cl	Various up to 4 wt %	Up to 260	>90	<10	Up to 2.0
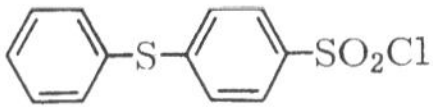 ⟨◯⟩-S-⟨◯⟩-SO$_2$Cl	1.2	240	>90	<10	0.56
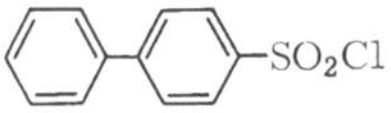 ⟨◯⟩-⟨◯⟩-SO$_2$Cl	1.3	250	95	100	—
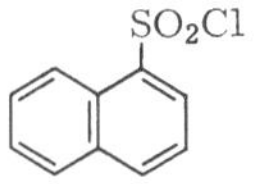 (naphthalene)-SO$_2$Cl	Various up to 40	Up to 290	90	0	<0.1
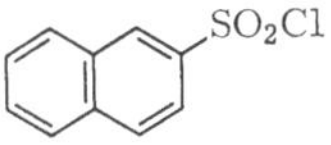 (naphthalene)-SO$_2$Cl	Various up to 40	Up to 290	90	0	<0.1

* As % of total product.
[1] Reduced viscosity = (t solution-t solvent)/t solvent for 1% solution in dimethyl formamide at 25° C.

6. Miscellaneous methods

The polymerization of cyclohexane 1.4-oxide has been described (87). A high melting, high molecular weight polymer was obtained. It is conceivable that this polymer could be dehydrogenated to the aromatic ether, however, no information is available on this point.

PRICE (74) has reported an attempted preparation of a polyphenylene ether via a pyrolysis of mercuriacetates. Only insoluble products were obtained.

GOLDEN (28) unsuccessfully attempted to prepare polymers by elimination of acetyl chloride and acetic anhydride from p-acetoxychlorobenzene and diacetoxybenzene, respectively. He was also unable to dehydrate hydroquinone. In the course of his study on the dehydration of phenols to aromatic ethers over thoria, BRINER (8) also attempted, unsuccessfully, to dehydrate the polyphenols pyrocatechol, resorcinol and hydroquinone.

Recently McNELIS has reported that oxidation of mesitol with activated manganese dioxide yields as the principal product the polyether resulting from removal of the p-methyl group. Formaldehyde was identified as a concomitant product (62).

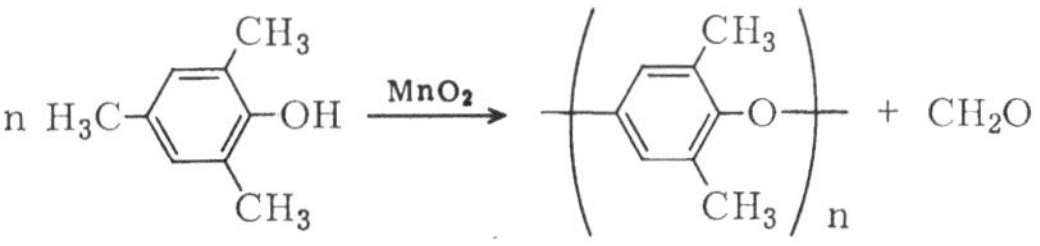

The polymers obtained were very low molecular weight.

III. Structure and properties of aromatic polyethers

1. Halogen substituted polyphenylene ethers

The polymers prepared by HUNTER from polyhalophenols were undoubtedly highly branched. He examined the decomposition of salts of 2.6-dibromo-4-chlorophenol, 2.6-dichloro-4-bromophenol and 2.6-diiodo-4-chlorophenol (44). By analysis of the resulting polymers he was able to determine qualitatively (Table 6) the order of reactivety of the halogens (I>Br>Cl) and furthermore established that considerable reaction occurs through the o-position.

Table 6

	2,6-Dibromo-4-chloro	2,6-Dichloro-4-bromo	2,6-Diiodo-4-chloro
Reacted o-	59.4	30.7	62.3
Reacted p	40.6	69.3	37.7
Ratio o/p	1/0.86	1/2.26	1/0.61

Thus on the basis of this data the polymers are far from being linear. It has been shown subsequently by BLANCHARD (5) that the intrinsic viscosities of polymers of this type are considerably lower than would be expected for linear polymers with these relatively high molecular weights. All of the polymers prepared by GOLDEN gave friable mouldings when they were compression moulded, however, this in itself is not definitive since the molecular weights were also low enough so that even if they were linear polymers they would probably be brittle materials. GOLDEN also found no evidence of crystallinity in these polymers which would be expected if they had any degree of linearity. The infrared spectra were complex and provided no quantitative structural data (28).

HAY also found that the oxidative polymerization of 2-chloro-6-methylphenol gave high molecular weight polymers with lower than expected intrinsic viscosities (35). Analysis of the polymers also showed loss of chlorine (10—15%) during the polymerization. Branch points such as (XIII) were assumed in this instance. In the case of the polyhalophenol

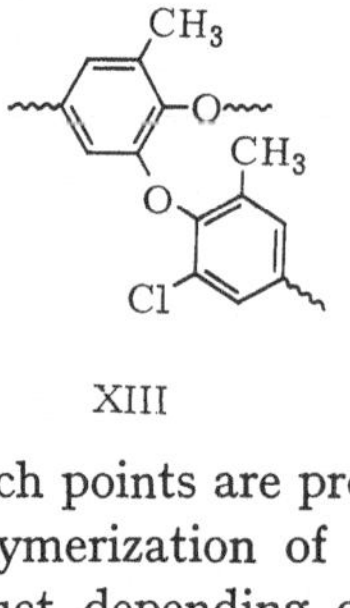

XIII

polymerizations similar branch points are probably present.

HUNTER found that polymerization of 2.4.6-triiodophenol yielded either a red or white product depending on the conditions (48). He

established the color in the former was due to quinone moieties by titration with hydrazine and measuring the nitrogen evolved. The colors were
similar to those of the diphenoquinones but the titrations showed it was
due to only a small portion of the molecule. It now appears that structures such as (XIV) could be readily incorporated into the polymers.

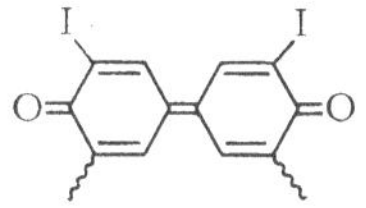

XIV

However, the recent work on polymerization of salts of 2.6-dichloro-4-
bromophenol under anhydrous conditions in the presence of aprotic
dipolar solvents apparently gives a linear high molecular weight polymer
as judged by the analytical data (see Experimental) and properties of
the polymer as summarized in the following tables (26).

Table 7

Flow temperature	300—310° C
Inherent viscosity	0.70
Density	1.466 g/cm

Table 8

Temperature, °C	Modulus, psi	Tensile strength, psi	Elongation %
23	400,000	12,600	9.4
100	384,000	8,600	8.4
150	352,000	6,000	7.8
200	300,000	2,700	14.0
225	260,000	200	136
250	240,000	50	268

2. Unsubstituted polyphenylene ethers

Poly-1.3-phenylene ethers prepared by BROWN (9, 10) have molecular
weights approaching 10,000 and softening temperatures in the range
60—70° C. They are therefore also below the molecular weight where
physical properties begin to optimize. Differential thermal analysis (DTA)
indicated a crystal melting point of 59° C and the infrared spectra were
consistent with the proposed structure. Hence these polymers are probably almost completely linear in structure. Thermogravimetric analysis
showed that major decomposition begins at about 500° C in nitrogen or
air, however some curing occurred during a 20 hour period in air at 150°C.
As BROWN has noted, this is probably due to oxidation reactions involving
the phenolic end group.

Very little data is given by STAMATOFF on the structure of properties
of poly-1.4-phenylene ether (75). There processing temperatures are in
the expected range for a high molecular weight linear polymer and it is
indicated that polymer separates out of the reaction mixture in the latter
stages of the reaction. This could be due to crystallization of the polymer.
It is indicated that fibers can be drawn from the melt and that the
polymer can be molded into shapes, however no physical properties of
the resulting products are given.

3. Alkyl-substituted polyphenylene ethers

The polymer prepared by oxidative coupling appears to be identical
with those prepared by PRICE by oxidative displacement of halogen.
Most of the reported data is on poly (2.6-dimethyl-1.4-phenylene ether)
(VII; $R_1=R_2=CH_3$). The infrared, ultraviolet and nuclear magnetic
resonance spectra are entirely consistent with the 1.4-polyether structure
(35, 74). It has also been demonstrated that this polymer (14) can be
crystallized. A plot of the logarithm of intrinsic viscosity vs. number
average molecular weight gives a straight line as shown in Fig. 1. This

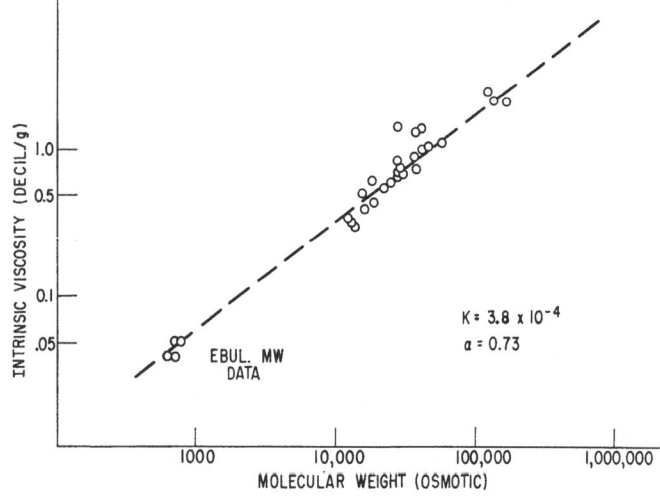

Fig. 1. Relationship of intrinsic viscosity to molecular weight ($\bar{M}_n$)

further confirms the essential linearity of the polymer. In addition several
polymers from other 2.6-disubstitution phenols have been described (35)
and the infrared spectra as well as viscosity-molecular weight data confirm
the 1.4-phenylene ether structure. Some of the physical properties of a
commercial grade of this polymer are summarized in the following table.

Table 9

Property	Average value
Specific gravity	1.06
Tensile strength	10,000—11,000 psi
Tensile modulus	2.6—3.8 $\times$ 10^5 psi
Elongation	6—7%
Dielectric Constant	2.58

The excellent overall properties of the polymer place it in the class of materials termed engineering thermoplastics.

KARASZ and O'REILLY (51) investigated the thermal properties of the polymer using a Model DSC-1 Differential Scanning Calorimeter (Perkin-Elmer Corporation). A glass transition temperature was observed at 205° C and a crystal melting-point at 267° C. The ratio of Tg/Tm is 0.92 which is the highest value that has been recorded for a polymer. On further heating, significant degradation began at about 457° C. This is some 60° higher than results that have been reported from a thermo-gravimetric study (42). In use in air at elevated temperatures, however, the methyl groups of this polymer undergo slow autoxidation reactions with concomitant loss of properties. As would be expected polymers of this type are unaffected by most aqueous acids and bases even at relatively high temperatures.

4. Polyethersulfones

The only data available to date on these polymers is the patent (50) and commercial literature. A large number of different polymers have been prepared (Table 1). The physical properties of the polymer from 4.4'-isopropylidenediphenol and 4.4'-dichlorodiphenyl sulfone are summarized in Table 10.

Table 10

Property	Value
Tensile strength	10,000 psi
Elongation	7%
Flex. modulus	4 $\times$ 10^5 psi
Tg °C	200

It is apparent that this polymer is extremely stable, hydrolytically, thermally and oxidatively, as well as being a tough and strong material which places it also in the category of materials known as engineering thermoplastics.

IV. Polymerization mechanisms

1. Ullmann condensation

In the past, various copper salts or copper metals have been commonly used as catalysts in the Ullmann condensation. No other metals or salts have been found to cause any appreciable catalysis. Despite the wide-spread use of this reaction in synthesis, little work on the mechanism of the reaction has been reported. Recent work by WEINGARTEN (84, 85) has led to the conclusion that the catalytic species in the reaction is a copper (I) salt and that the reaction is best described as a typical nucleophilic aromatic substitution reaction in which the mobilities of the halogens are in the expected order I~Br>Cl≫F. WEINGARTEN's work was carried out primarily in diglyme solvent. He found that the rates of reactions were considerably faster in unpurified diglyme due to catalysis by autoxidation products of the diglyme. Many esters of ethylene glycol were subsequently found to cause this rate enhancement. Under these optimum conditions the condensation has been found to take place at an appreciable rate as 100° C in contrast to the usual temperatures which have been employed in the neighborhood of 200° C or higher.

BACON and HILL (3) have found that copper (I) oxide can be conveniently used as a source of the copper catalyst. They performed the condensation simply by heating the phenol and aromatic halide in the presence of the oxide in the aprotic dipolar solvents, collidine, pyridine, dimethyl sulfoxide or dimethyl formamide. Neither of these groups appear to have used these milder conditions for the preparation of polymers.

The preparation of the related high molecular weight poly-1.4-phenylene sulfide has been accomplished by heating p-bromothio-phenolate salts in pyridine at 250° C (57). The commercially available polyethersulfones are reported to be prepared by condensation of 4.4'-dichlorodiphenyl sulfone with salts of biphenols in solvents such as dimethylsulfoxide at 150° C. The work of BACON and HILL would suggest that both of these reactions might be carried out at considerably lower temperatures with copper (I) salts as catalysts. In addition, it has been demonstrated that copper (I) acetylides react quantitatively with aromatic iodides to yield tolanes (15, 77); therefore this reaction should also be the basis for a similar polymer forming reaction.

2. Oxidative polymerization

Phenols such as 2.6-dimethylphenol are converted rapidly and in high yield to high molecular weight polymers at room temperature with oxygen in the presence of amine complexes of copper salts as catalyst. Much of the work described in the literature has been performed with copper (I) chloride as catalyst and pyridine as ligand and solvent. Other amines, primary, secondary or tertiary can be used as ligands for the catalyst. Autoxidation of copper (I) chloride in pyridine results in the absorption of one mole of oxygen per four moles of copper(I) chloride which would indicate the following reaction.

$$4 \text{ CuCl} + O_2 \rightarrow \text{Cu(OH)Cl} + 2 \text{ H}_2\text{O}$$

However, on attempted isolation, the product obtained is a complex, partially amorphous, basic copper (II) oxychloride intimately associated with bis pyridine copper (II) chloride (5, 25).

Copper (II) salts have been found to be inactive as catalysts for the reaction with the exception of the copper (II) carboxylates which are considerably less reactive. In addition, the polymerization of 2.6-xylenol in pyridine with copper (II) acetate as catalyst appears to terminate before high molecular weight polymers are formed. However, treatment of an amine complex of a copper (II) salt with an equivalent of a strong gives the active catalyst. Similarly, although copper (II) hydroxide in pyridine is inactive as a catalyst, treatment with an equivalent of hydrogen chloride generates the active catalyst. Hence it can be concluded that the active catalyst is a basic salt (XV).

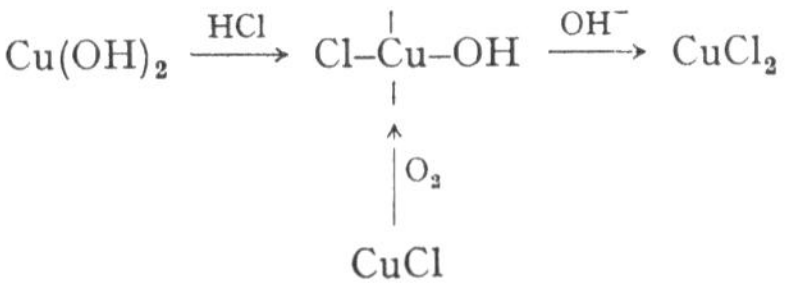

$$\text{Cu(OH)}_2 \xrightarrow{\text{HCl}} \text{Cl–}\overset{|}{\underset{|}{\text{Cu}}}\text{–OH} \xrightarrow{\text{OH}^-} \text{CuCl}_2$$

$$\overset{\wedge}{\underset{|}{}}O_2$$

CuCl

XV

Endres has studied the effect on the reaction of varying the pyridine to copper ratio in the catalyst (Table 11) (22).

Table 11. *Effects of varying pyridine concentration* [a]

Pyridine M	Ligand ratio, N/Cu	$R_{max} \times 10^3$ mole 1^{-1} min^{-1}	Oxygen absorbed %	Fractional Yields		Intrinsic viscosity decil. g^{-1}
				fc—o	fc—c	
0.0033	0.67	0.091	107	0.072	0.56	[c]
.0050	1.0	.206	105	.16	.49	[c]
.0100	2.0	.662	96.5	.40	.34	0.086
.0150	3.0	1.26	100	.51	.26	.097
.050	10	5.20	99	.75	.10	.17
.50	100	11.4	98.5	.86	0	.49
2.79	558	7.70	99	.82	0	.725
9.00	1800	1.30	108	.785	0	.71
9.00	1800	1.33	111	.79	0	.76
12.1 [b]	2420	0.666	109	.80	0	.94

[a] Conditions: 2.6-Dimethylphenol 0.2 M; copper (I) chloride 0.005 M; o-dichlorobenzene solvent; 30°. At ligand ratios 0.2 to 100, anhydrous magnesium sulfate was also present at 0.2 mole/l.

[b] Pyridine solvent.

[c] Insufficient sample for determination.

At the lowest ratio (0.67) at which a homogeneous solution (o-dichloro-benzene solvent) can be obtained a slow reaction leads predominantly to the carbon-carbon coupled product, 3.3',5,5'-tetra-methyldiphenoqui-none. When the ligand ratio is increased, C-O coupling to yield polymer gradually increases and at a ten to one ratio only minor amounts of diphenoquinone are formed. In contrast ENDRES has found (Fig. 2) that

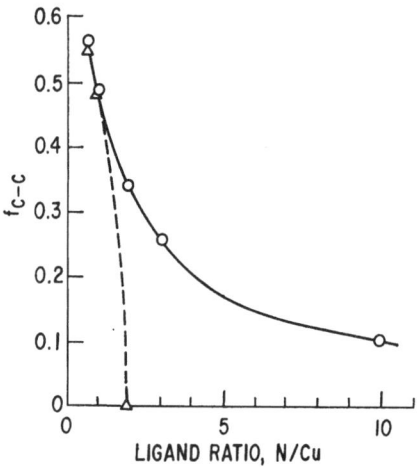

Fig. 2. Effect of nitrogen to copper ratio (o, pyridine; Δ, N,N,N',N'-tetramethylethylene diamine)

when the strongly basic bidentate amine, N,N,N',N'-tetramethylethylene diamine is used as ligand a sharp break occurs at 2 N/Cu above which almost exclusive C-O coupling occurs. Hence we can probably conclude that the active catalyst has two amino nitrogens which are probably trans to each other in the case of monodentate amines but necessarily cis with bidentate amines. The catalyst can then be represented as (XVI).

$$
\begin{array}{ccc}
R_3 & & R_3 \\
N & & N \\
| & & | \\
Cl\text{-}Cu\text{-}OH & or & R_3N\text{-}Cu\text{-}OH \\
| & & | \\
N & & Cl \\
R_3 & &
\end{array}
$$

XVI

At low ligand ratios using a mondentate amine, more than one species is probably present in solution and only at higher ligand ratios is most of the copper present in the active catalyst form. Since a bidentate amine like N,N,N',N'-tetramethylethylene diamine forms a stable chelate in other systems, this may account for the reason that only a molar amount is necessary in this case. The first step in the reaction then appears to be (25)

$$-\overset{|}{\underset{|}{C}}uOH + HOAr \rightarrow -\overset{|}{\underset{|}{C}}uOAr + H_2O$$

ENDRES found that increasing the size of the ligand coordinated to copper favored C-C coupling (*22*). HAY has reported that the oxidation of *o*-cresol with pyridine as ligand for the catalyst yields very low-molecular weight materials. However if bulky amines are used as ligands, high molecular weight although not completely linear polymers are obtained. It was assumed that the bulky ligand in this case physically blocked most of the reaction at the open *o*-position (*38*).

In this copper-amine complex, electron transfer from oxygen to copper would give a phenoxy radical which, since the principal reaction of free phenoxy radicals appears to be C-C coupling, in this case must remain bound to the catalyst. Coupling of two of these radicals would then give dimer.

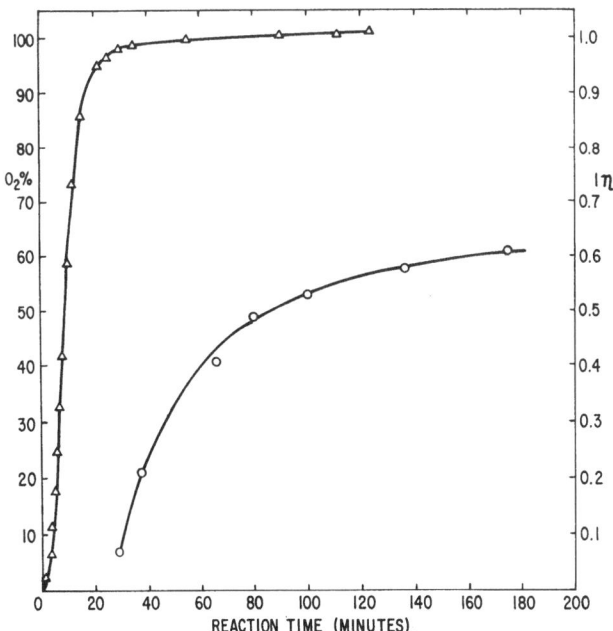

A priori it might have been expected that the propagation reaction involved simply the addition of monomer units in a similar fashion to the end of the growing chain. ENDRES studied the course of this reaction by sampling at various times and analysis of the products (*23*). His results

RELATION BETWEEN OXYGEN UPTAKE
AND VISCOSITY OF POLYMER PRODUCED

Fig. 3. Relation between oxygen uptake and viscosity of polymer produced

showed that high molecular weight polymer is only formed in the latter stages of the reaction. This is best illustrated by comparing the viscosity of the polymer at a given time with percent completion of reaction as measured by oxygen absorption. A typical result is shown in Fig. 3. It is apparent that mechanistically the polymerization reaction behaves as a typical polycondensation reaction rather than as an addition-type polymerization. In confirmation of this, it was also found by ENDRES and KWIATEK that dimer (XVII) and trimer (XVIII), prepared by independent methods, as well as low molecular weight oligomer, isolated from a polymerization reaction at an early stage, could be polymerized in high

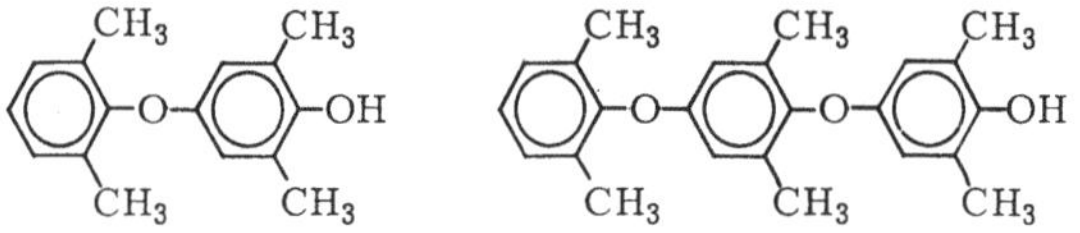

XVII   XVIII

yield, at comparable rates, to high molecular weight polymers (23, 56). It was also found that if the hydroxyl group in these low molecular weight species was protected (by methyl groups) they were then completely unreactive in a reaction mixture, in the presence or absence of monomer.

To explain this unusual result, two basically different mechanisms were proposed (24). The first, the uncoupled electron mechanism assumed that oligomeric phenoxy radicals combined by a head-to-tail coupling reaction. This implied then that a resonance structure could be written

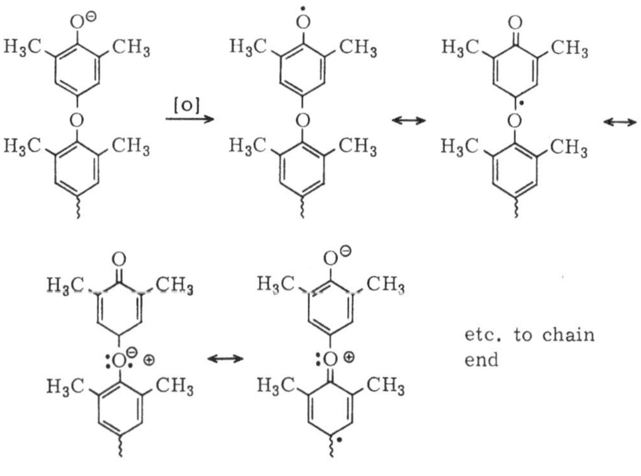

etc. to chain end

that would allow the odd electron on an oligomeric phenoxy radical to "see" the end of the chain. PRICE found that in the oxidative polymerization of 2.6-xylenol-3-H^3 and 2.6-xylenol-4-H^3 16% of the label in the former was lost and 23% of the label in the latter was retained (*14*). To explain this he proposed the following extension of the uncoupled elec-

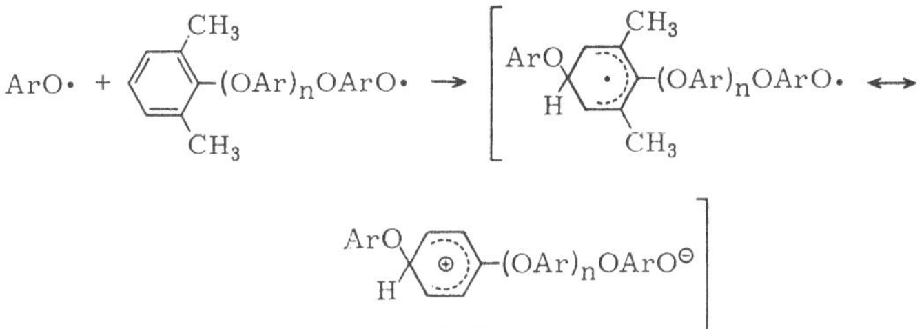

XIX

tron mechanism. He proposed that the phenomium intermediate (XIX) could lose the 4-proton directly or undergo facile proton migration.

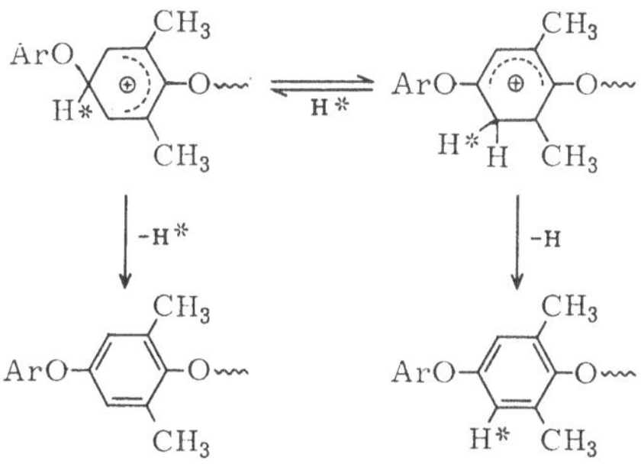

The other mechanism proposed also involved coupling of oligomeric radicals to give quinol ethers as intermediates and is illustrated as follows for dimer.

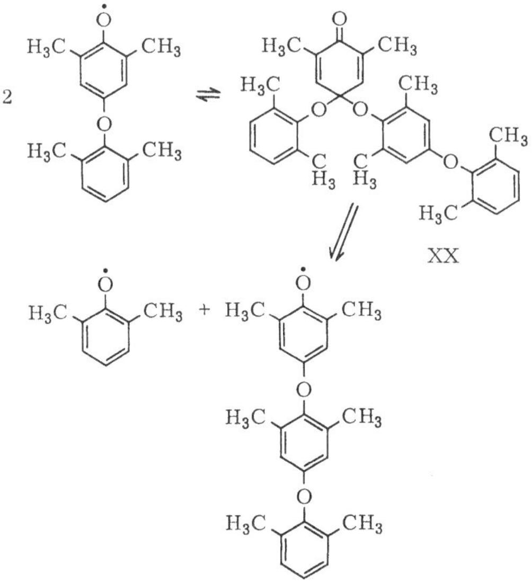

XX

From this point two alternatives were proposed. The first has been called the quinol ether rearrangement and involves a rearrangement of the intermediate (XX) of the benzidine type (49) or the quinamine rearrangement recently discussed by MILLER (64). The second alternative proposed involved quinol ether equilibration. The unstable quinol ether intermediate (XX) could dissociate either to give the aryloxy radicals started with or to give two new aryloxy radicals, in this case derived from monomer and trimer. These two could then couple to give tetramer. A similar mechanism was also proposed by MÜLLER (65) to explain the products obtained in the oxidation of polyhalophenols. In order to distinguish between these two mechanisms, the dimer has been oxidized under polymerization conditions and the reaction mixture analyzed at intermediate stages. The results obtained by COOPER (18) are shown in Fig. 4. If the quinol ether rearrangement occurred only even numbered

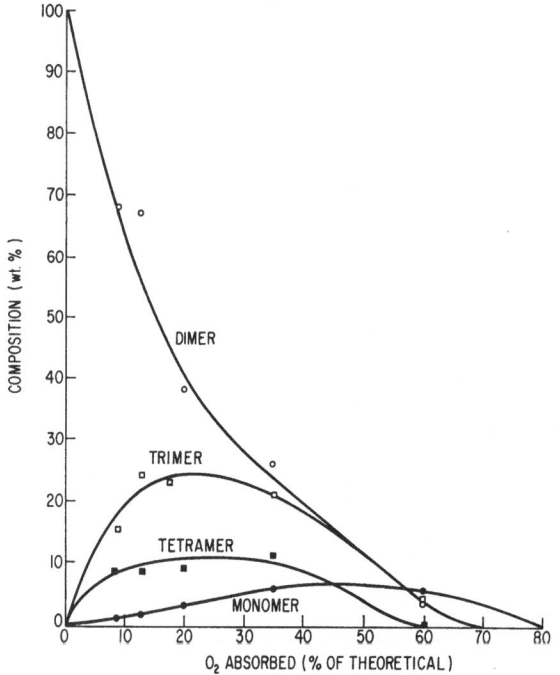

Fig. 4. Composition of product from xylenol dimer as a function of extent of oxidation

oligomers would be present; since monomer and trimer are shown to be present, the quinol ether equilibration must take place under the conditions used. Similar results have been demonstrated by other workers (12, 19). PRICE had attempted to equilibrate high polymer ($[\eta] = 0.93$) with monomer and isolated polymer with $[\eta] = 0.76$. He cited this as evidence that equilibration did not occur; however, it has been pointed out that this drop in intrinsic viscosity is equivalent to a change in degree of polymerization from 535 to 340, indicating considerable redistribution must have occurred (*19*). The same conclusion was reached by McNELIS (*61*) who also pointed out that the loss of tritium in the 3-position or retention of tritium in the 4-position was, in fact, in accord with the quinol ether mechanism. He pointed out that in this case exchange can take place since the quinol ether intermediate (XX) would tautomerize to a phenol via a protonation step during which the exchange can take place.

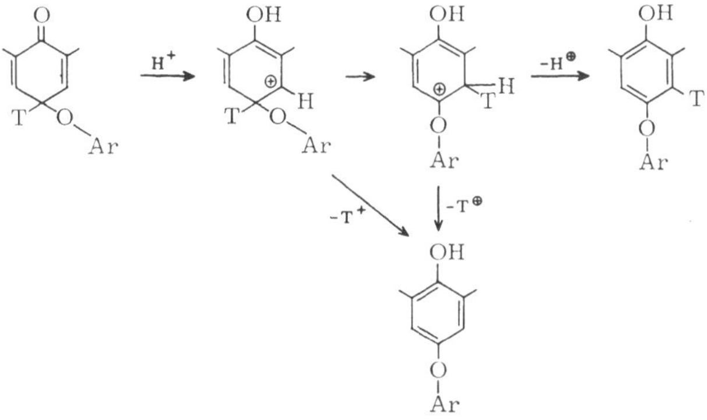

MIJS and coworkers (*63*) have oxidized a series of *p*-phenoxyphenols (XXI) with

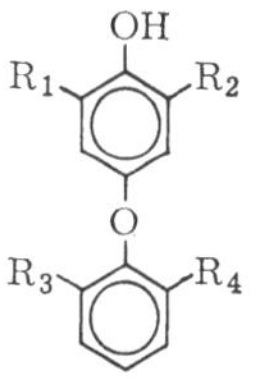

(a) $R_1=R_2=R_3=R_4=CH_3$
(b) $R_1=R_2=R_3=R_4=H$
(c) $R_1=CH_3$, $R_2=R_3=R_4=H$
(d) $R_3=CH_3$, $R_1=R_2=R_4=H$
(e) $R_1=R_2=CH_3$, $R_3=R_4=H$
(f) $R_1=R_2=H$, $R_3=R_4=CH_3$

XXI

varying numbers of methyl substituents. From XXI (a) at 40° C, as noted by COOPER, monomer and trimer were the first main reaction products. Oxidation of XXI (b) at 40° gave only even numbered oligomers as products but at 80° C monomer and trimer were also formed. At 5° C no monomers and trimers were detectable from XXI (c) and XXI (d) but they were present when the oxidation was performed at higher temperatures. The other dimers gave monomers and trimers as products. The main reaction path for XXI (b), XXI (c), and XXI (d) must then be the quinone ketal rearrangement and presumably because of steric hindrance in the other three cases, homolytic fission of the intermediate quinone ketals occur.

Further evidence against the "uncoupled electron" mechanism was obtained by analysis of the products. Oxidation of **XXI** (c) gave tetramer (**XXII**),

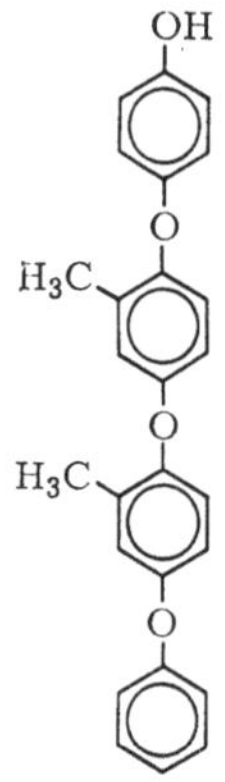

XXII

which could not have been produced via the latter route.

It has also been shown that xylenol dimer (**XVII**) can be redistributed with other phenols in the presence of initiators (*6, 63*). BOLON (*6*) has arranged a number of phenols into a relative reactivities scale (Table 12) by reacting xylenol dimer with two different phenols and measuring the relative amounts of the two new dimers (**XXIII, XXIV**).

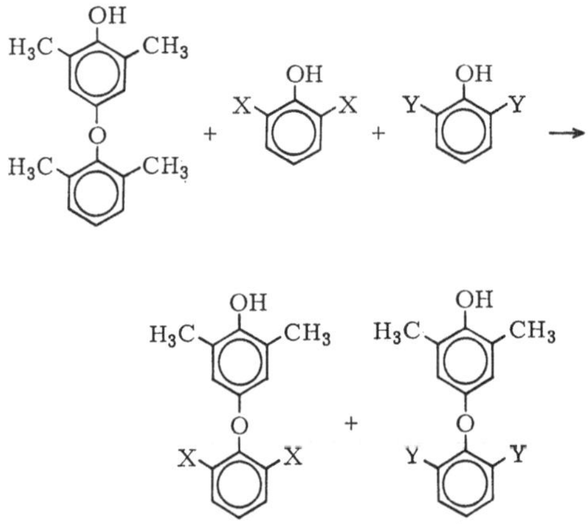

XXIII   XXIV   + others

WHITE (*86*) has extended the work of COOPER on equilibration of monomer with polymer and used it for the preparative synthesis of low molecular weight oligomers. From the reaction of 1000 g of polymer with 1000 g of 2.6-xylenol using tetramethyl diphenoquinone as initiator, he obtained (as acetates) 293 g of dimer, 220 g of trimer and 35 g of tetramer. Previous methods for the preparation of these materials involved many tedious steps. Mixed dimers can also be prepared by using a phenol different from

Table 12

Phenol	Relative Co-redistribution
4-methoxy-2.6-dimethyl-	100
2.4.6-trimethyl-	29
4-methoxy-	23
4-chloro-2.6-dimethyl-	10
4-t-butyl-	10
2.6-diisopropyl-	7.5
4-ethyl-	3.5
phenol	0

that incorporated in the polymer. In this case, as predicted from the quinol ether mechanism, in the dimer the non-phenolic end comes from the phenol and the phenolic end from the polymer, e.g.,

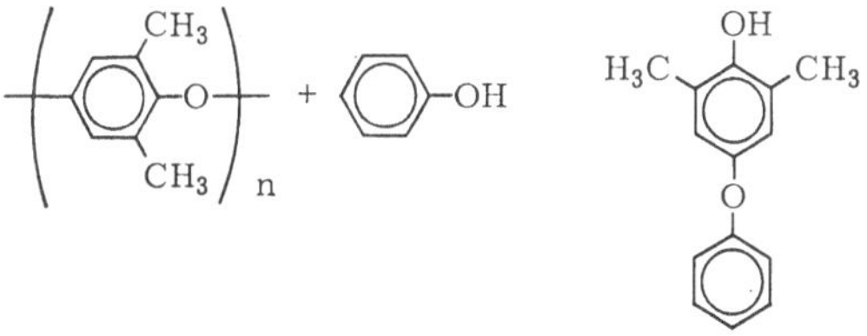

3. Oxidative displacement of halogen

Polymerization by oxidative displacement of halogen differs from oxidative polymerization in that less than stoichiometric amounts of oxidizing agent are necessary. Optimum yields of high molecular weight polymer are obtained with about 10 mole per cent of initiator ($K_3Fe(CN)_6$, PbO_2, Ag_2O, etc.). Hence a free radical chain must be invoked to explain the polymerization and PRICE has proposed the following mechanism (74).

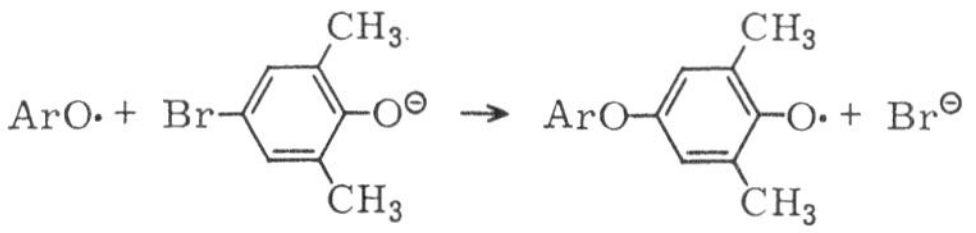

It is interesting to note, however, that HAMILTON and BLANCHARD (34) have prepared the brominated dimer (XXV) and they find that it

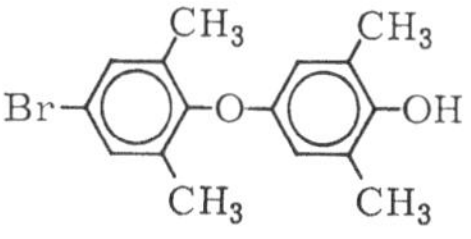

XXV

polymerizes readily under the same conditions as 4-bromo-2.6-xylenol. This can only be explained (in the light of the discussion in the preceding section) by assuming the following quinol ether intermediate (XXVI).

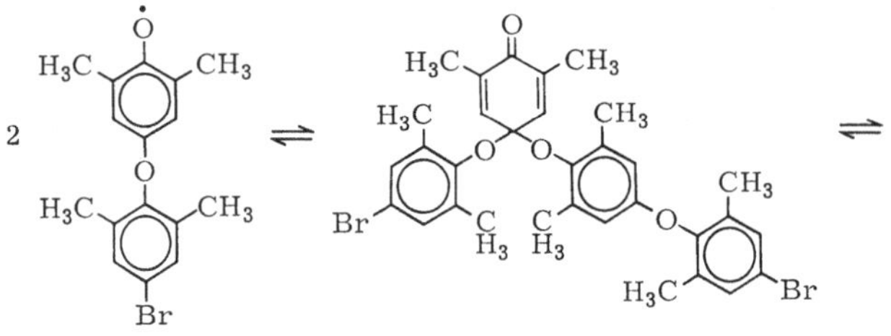

XXVI

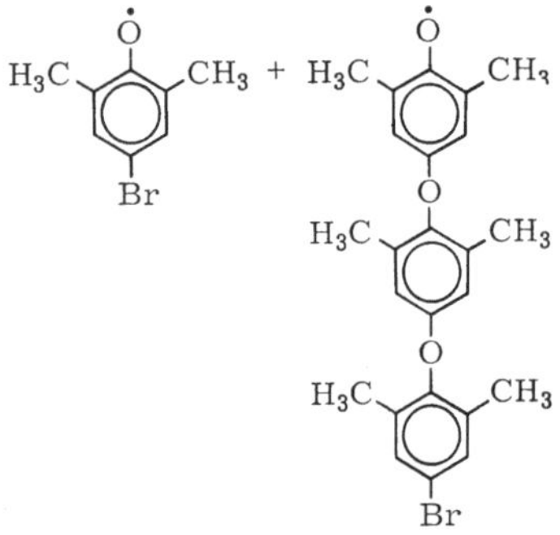

These monomeric radicals so formed would then fit into the scheme as proposed by PRICE.

BECKER (4) has found that oxidation of 2.6-di-*t*-butyl-4-bromo-phenol in the presence of pentachlorophenol gives the cyclohexadienone (XXVIII).

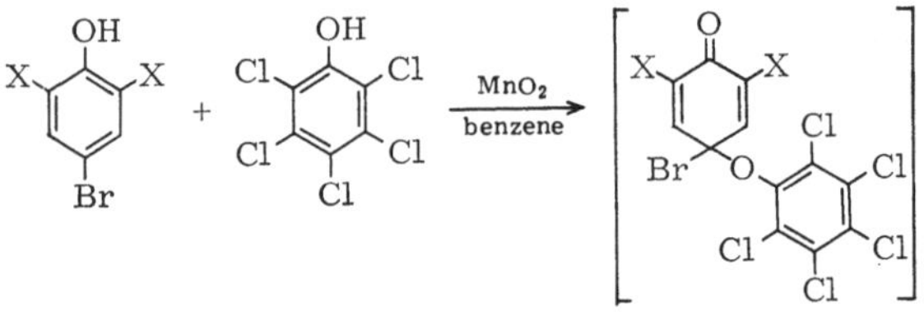

XXVII

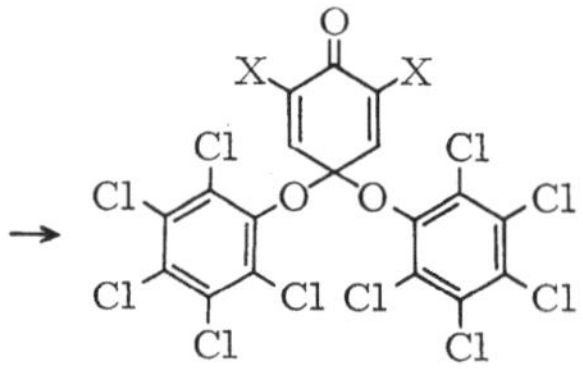

XXVIII

It was also shown that this aromatic quinone ketal dissociates at room temperature as follows

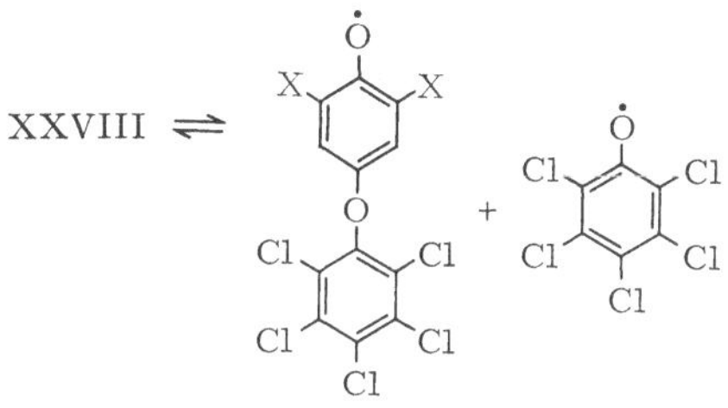

XXVIII ⇌

It is reasonable to assume that the polymerization of halophenols as described by PRICE proceeds through similar intermediates. The formation of 2.6-dimethylbenzoquinone (67) as a byproduct could then reasonably result from hydrolysis of the intermediate (XXVII). This then would also explain the necessity of using large amounts (~ 10 mole percent) of initiator in the polymerization since initiator would be used up in this reaction.

The polymerization of trihalophenols can be described by a similar mechanism. Since the reaction proceeds through the o-position as well as the p-position intermediates such as (XXIX) must be assumed. It is

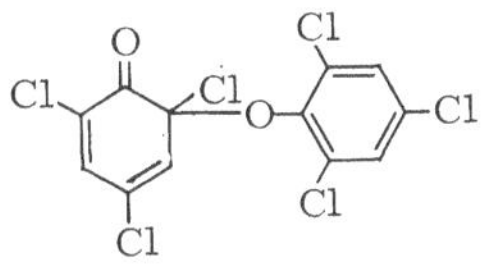

XXIX

striking that when the reaction is run under anhydrous conditions (26) reaction occurs almost exclusively through the p-position and in addition the yields are quantitative. It is noteworthy that under these conditions much less initiator is required. Side reactions involving hydrolytic cleavage of intermediates could not occur here. It would also be expected that under anhydrous conditions in the presence of aprotic, dipolar solvents the formation of intermediates similar to (XXVIII) from (XXVII) would be facilitated.

Bibliography

1. AUWERS, K., and T. MARKOVITS: Chem. Ber. 38, 226 (1905).
2. —, and G. WITTIG: Chem. Ber. 57B, 1270 (1924).
3. BACON, R. G. R., and H. A. O. HILL: J. Chem. Soc. 1108 (1964).
4. BECKER, H.-D.: J. Org. Chem. 29, 3068 (1964).
5. BLANCHARD, H. S., H. L. FINKBEINER, and G. A. RUSSELL: J. Pol. Sci. 58, 469 (1962).
6. BOLON, D. A.: Polymer Preprints, Vol. 7, No. I, Am. Chem. Soc. Mtg., Phoenix, Arizona, January, 1966, p. 173.

7. BRINER, E., J. BRON-STALET, and H. PAILLARD: Helv. Chim. Acta **15**, 619 (1932).
8. —, and A. A. BRON: Helv. Chim. Acta **15**, 1234 (1932).
9. BROWN, G. P., and A. GOLDMAN: Polymer Preprints, **5**, No. 2, Am. Chem. Soc. Mtg. New York, N. Y., September, 1963, p. 39.
10. — Polymer Preprints, **5**, No. 1, Am. Chem. Soc. Mtg. Philadelphia, Pennsylvania, April, 1964, p. 195.
11. — Private communication.
12. BUSSINK, J., O. E. VON LOHUIZEN, J. JULDER, and L. VOLLBRACHT: Private communication.
13. BUTTE, W. A., N. S. CHU, and C. C. PRICE: Macromolecular Syntheses, Vol. I, p. 76. New York: John Wiley and Sons, Inc. 1963.
14. —, C. C. PRICE, and R. E. HUGHES: J. Polymer Sci. **61**, 528 (1962).
15. CASTRO, C. E., and R. D. STEPHENS: J. Org. Chem. **28**, 2163 (1963).
16. CLEMO, G. R.: J. Chem. Soc. 1265 (1931).
17. —, and R. SPENCE: J. Chem. Soc. 2811 (1928).
18. COOPER, G. D., H. S. BLANCHARD, G. F. ENDRES, and H. L. FINKBEINER: J. Am. Chem. Soc. **87**, 3996 (1965).
19. —, A. R. GILBERT, and H. L. FINKBEINER: J. Org. Chem. **31**, (in press).
20. CUDBY, M. E. A., R. G. FEASEY, B. E. JENNINGS, M. E. B. JONES, and J. B. ROSE: Polymer **6**, 589 (1965).
21. DEWAR, M. J. S., and A. N. JAMES: J. Chem. Soc. 917 (1958).
22. ENDRES, G. F., A. S. HAY, and J. W. EUSTANCE: J. Org. Chem. **28**, 1300 (1963).
23. —, and J. KWIATEK: J. Polymer Sci. **58**, 593 (1962).
24. FINKBEINER, H. L., G. F. ENDRES, H. S. BLANCHARD, and J. W. EUSTANCE: Soc. Plastics Eng. Trans. **2**, 112 (1962).
25. —, A. S. HAY, H. S. BLANCHARD, and G. F. ENDRES: J. Org. Chem. **31**, 549 (1966).
26. *French Patent* 1,403,987 (May 17, 1965).
27. GLADSTONE, J. H., and A. TRIBE: J. Chem. Soc. **41**, 5 (1882).
28. GOLDEN, J. H.: Soc. Chem. Ind. (London) Monogr. No. 13, 231 (1961).
29. HALE, W. J., and E. C. BRITTON: U. S. Patent 1,737,841 (Dec. 3, 1929).
30. — U. S. Patent 1,737,842 (December 13, 1929).
31. — U. S. Patent 1,806,798 (May 26, 1931).
32. — U. S. Patent 1,882,824 (October 18, 1933).
33. — U. S. Patent 1,925,321 (September 5, 1933).
34. HAMILTON, S. B., and H. S. BLANCHARD: Private communication.
35. HAY, A. S.: J. Polymer Sci. **58**, 581 (1962).
36. —, H. S. BLANCHARD, G. F. ENDRES, and J. W. EUSTANCE: J. Am. Chem. Soc. **81**, 6335 (1959).
37. — — — Macromolecular Syntheses, Vol. I, p. 75. New York: John Wiley and Sons, Inc. 1963.
38. —, and G. F. ENDRES: Polymer Letters **3**, 887 (1965).
39. HAYNES, C. G., A. H. TURNER, and W. A. WATERS: J. Chem. Soc. 2823 (1956).
40. HEDAYATULLAH, M., and L. DENIVILLE: Compt. rend. **254**, 2369 (1962).
41. HOFFMEISTER, W.: Ann. Chem. **159**, 191 (1871).
42. HOYT, H. E., B. D. HALPERN, K. C. TSOU, M. BODNAR, and W. TANAR: J. Appl. Pol. Sci. **8**, 1633 (1964).
43. HUNTER, W. H., and M. A. DAHLEN: J. Am. Chem. Soc. **54**, 2456 (1932).
44. —, and F. E. JOYCE: J. Am. Chem. Soc. **39**, 2640 (1917).
45. —, A. O. OLSEN, and E. A. DANIELS: J. Am. Chem. Soc. **38**, 1761 (1916).
46. —, and L. M. SEYFRIED: J. Am. Chem. Soc. **43**, 151 (1921).

47. HUNTER, W. H., and R. B. WHITNEY: J. Am. Chem. Soc. 54, 1167 (1932).
48. —, and G. H. WOOLLETT: J. Am. Chem. Soc. 43, 131, 135 (1921).
49. INGOLD, C. K., and H. V. KIDD: J. Chem. Soc. 984 (1933).
50. JOHNSON, R. N., and A. G. FARNHAM: Dutch Patent Application 6,408,130.
51. KARASZ, F. E., and J. M. O'REILLY: Polymer Letters 3, 561 (1965).
52. KUNITAKE, T., and C. C. PRICE: J. Am. Chem. Soc. 85, 761 (1963).
53. KURIAN, C. J., and C. C. PRICE: J. Polymer Sci. 64, 267 (1961).
54. KWIATEK, J.: Private communication.
55. — U. S. Patent, 3,133,899 (May 19, 1964).
56. — U. S. Patent 3,134,753 (May 26, 1964).
57. LENZ, R. W., C. E. HANDLOVITS, and H. A. SMITH: J. Polymer Sci. 58, 351 (1962).
58. LINDGREN, B. O.: Acta Chem. Scand. 14, 1203 (1960).
59. — Acta Chem. Scand. 14, 2089 (1960).
60. MCNELIS, E. J.: U. S. Patent 3,220,979 (November 30, 1965).
61. — J. Org. Chem. 31, 1255 (1966).
62. — J. Am. Chem. Soc. 88, 1074 (1966).
63. MIJS, W. J., O. E. VON LOHUIZEN, J. BUSSINK, and L. VOLLBRACHT: J. Am. Chem. Soc. 87, in press (1966).
64. MILLER, B.: J. Am. Chem. Soc. 86, 1127 (1964).
65. MÜLLER, E., A. RIEKER, and W. BECKERT: Z. Naturforschg. 176, 567 (1962).
66. OGATO, Y., and T. MORIMOTO: Tetrahedron 21, 2791 (1965).
67. PRICE, C. C., and N. S. CHU: J. Polymer Sci. 61, 135 (1962).
68. SABATIER, P., and A. MAILHE: Compt. rend. 151, 493 (1910).
69. — Compt. rend. 155, 261 (1912).
70. — Bull. Soc. Chem. Fr. [4] 11, 843 (1912).
71. — Compt. rend. 158, 611 (1914).
72. SAX, K. J., W. S. SAARI, C. L. MAHONEY, and J. M. GORDON: J. Org. Chem. 25, 1590 (1960).
73. SHORYGIN, P. P., I. S. KIZBER, V. I. ISAGULYANTS, and E. K. SMULYANINOVA: Khim. Referat. Zhur., No. 4, 116 (1940).[CA 36, 3792 (1940)].
74. STAFFIN, G. D., and C. C. PRICE: J. Am. Chem. Soc. 82, 3632 (1960).
75. STAMATOFF, G. S.: U. S. Patent 3,228,910 (January 11, 1966).
76. — U. S. Patent 3,236,807 (February 22, 1966).
77. STEPHENS, R. D., and C. E. CASTRO: J. Org. Chem. 28, 3313 (1963).
78. STILLE, J. K.: C. S. Marvel Symposium, University of Arizona, Dec. 27, 1961.
79. SÜS, O., K. MÖLLER, and H. HEISS: Ann. Chem. 598, 123 (1956).
80. TORREY, H. A., and W. H. HUNTER: J. Am. Chem. Soc. 33, 194 (1911).
81. ULLMANN, F., and A. STEIN: Chem. Ber. 39, 623 (1906).
82. UNGADE, H. E.: Chem. Rev. 38, 405 (1946).
83. WANG, C.: Proc. Chem. Soc. 309 (1961).
84. WEINGARTEN, H.: J. Org. Chem. 29, 977 (1964).
85. — J. Org. Chem. 29, 3624 (1964).
86. WHITE, D. M.: Polymer Preprints, Vol. 7, No. I, Am. Chem. Soc. Mtg., Phoenix, Arizona, January, 1966, p. 178.
87. WITTBECKER, E. L., H. K. HALL JR., and T. W. CAMPBELL: J. Am. Chem. Soc. 82, 1218 (1960).
88. WOOLLETT, G. H.: J. Am. Chem. Soc. 38, 2474 (1916).

Received July 21, 1966

Adv. Polymer Sci., Vol. 4, pp. 528—590 (1967)

Polytetrahydrofuran

By

P. Dreyfuss and M. P. Dreyfuss

B. F. Goodrich Company, Research Center
Brecksville, Ohio 44141

With 34 Figures

Table of Contents

I. Introduction

Summaries of the work in tetrahydrofuran polymerizations (*1, 2, 3*) have appeared as late as 1963. However, in the last four to five years the number of publications has been so numerous and the advances in the understanding of tetrahydrofuran polymerizations have been so rapid that it is worth reviewing again at this time. New catalysts have been reported, significant studies with old catalysts have been made, and a number of papers on the physical properties of polytetrahydrofuran have appeared. We will emphasize this new work and attempt to point out some areas where new investigations or a reinvestigation of earlier studies would be helpful.

A. Nomenclature

The monomer, commonly known as tetrahydrofuran, has the systematic name tetramethylene oxide or 1,4-epoxybutane. The polymer derived from it is as often called polytetramethylene oxide as polytetrahydrofuran. In accordance with Chemical Abstracts practice, we have chosen to use the names tetrahydrofuran (THF) and polytetrahydrofuran (PTHF).

B. Polymerizability of cyclic ethers

The polymerization of one of the three membered cyclic ethers was first carried out about a century ago (*4*), but the polymerization of THF was not accomplished until the late 1930's (*5*). This is not surprising when the thermodynamics of ring opening polymerizations is considered. Dainton and Ivin (*6*) have estimated the free energy of polymerization, ΔF_p, as a function of ring size. Usually the heat of polymerization, ΔH_p, is the main contributor to ΔF_p. The differences in ΔH_p are caused mainly by changes in ring strain and crowding of eclipsed adjacent hydrogen atoms. As shown in Table 1, the polymerization of three and

Table 1. *Heat of polymerization of cyclic ethers* (*6*)

Compound	$-\Delta H_p$ kcal. mole^{-1}	Polymerizability when substituted
Ethylene oxide	22.6	$+$
Oxetane	19.3	$+$
THF	3.5	$-$
Tetrahydropyran	-1.3	$-$

four membered cyclic oxides has a large negative ΔH_p and polymerization occurs readily, probably largely due to ring strain. THF also has a negative ΔH_p, but it is much smaller and THF polymerizes less readily than the three and four membered rings. Its polymerizability probably results from repulsions of eclipsed hydrogens (7). When the THF ring is substituted, ΔF_p becomes positive (6) and substituted THF's have not been reported to polymerize*. This is probably an entropy effect similar to the "gem-dimethyl" effect in organic chemistry (7). Tetrahydropyran which also has a positive ΔF_p does not polymerize either.

C. Catalysts for THF polymerization

THF polymerizes only by a cationic mechanism. Therefore, all of the catalysts are of the strong acid or Lewis acid type or of salts derived from them.

Polymerization of THF was first observed by MEERWEIN and his coworkers (8) and was studied extensively by this group in the 1930's and 1940's. However, this work did not become generally known until after World War II and even then it was available only in the form of microfilmed reports (9). It was not until 1960, with the publication of Meerwein's review (3), that the scope of this excellent work became generally available. The catalysts used by this group are basically of the trialkyl oxonium ion type, either preformed or generated in situ. MEERWEIN classified combinations which generate oxonium ions into three broad groups:

1) Combinations of metal halides (e. g. $FeCl_3$, $AlCl_3$) with compounds containing an active halogen atom (e. g. α-chloro-dimethyl ether, benzyl chloride, 2,3-dichloro THF).

2) Unsaturated tertiary oxonium salts, $[RC(OR')_2]^+X^-$, where $R=H$, CH_3, C_2H_5; $R'=CH_3$, C_2H_5; and $X^- = BF_4^-$, $SbCl_6^-$**. These salts can be preformed or formed in situ from the corresponding orthoester and the metal halide.

3) Other catalysts. These include "complex inorganic acids" (such as $HClO_4$, HBF_4, HSO_3F, H_2SnCl_6) frequently formed in situ, and the "acylium" salts formed from Lewis acids (e. g. BF_3, $AlCl_3$) and acylating agents (such as acetyl chloride, acetic anhydride).

It must be emphasized that the above catalysts differ considerably in their effectiveness. Some give solid polymers but most of the combinations give only liquid polymers. In addition the conversions realized vary widely.

Related catalysts have continued to be reported. These include SiF_4 (11), SeF_6 (47), WCl_6 (12), $NbCl_5$ (12), and $TaCl_5$ (12). But by far the most noteworthy of the Lewis acid class of catalysts is PF_5 (13, 14),

---

* 7-Oxabicyclo[2:2:1] heptanes which polymerize readily may be considered disubstituted THF's, but due to the highly strained ring system, they are obviously a special case.

** For a discussion of the chemistry and preparation of these salts see MEERWEIN (10).

which for the first time led to a PTHF of a molecular weight of several hundred thousand.

More recently there have been reports of a number of catalyst systems involving alkyl aluminium compounds (15, 16) (such as AlEt$_3$, AlEt$_2$Cl, AlEtCl$_2$) used together with cocatalsyts (usually water or epichlorohydrin).

Certain related complex ions have been successfully applied in recent studies of THF polymerizations. These include Ph$_3$C$^+$SbCl$_6^-$ (17, 18), Tropylium$^+$SbCl$_6^-$ (19), Ph$_3$C$^+$PF$_6^-$ (20), and 4-ClC$_6$H$_4$N$_2^+$PF$_6^-$ (21). These catalysts also give high molecular weight PTHF.

II. Ceiling temperature

It was realized very early that the polymerization of THF is reversible and that limiting conversions are reached at a given temperature. In the meantime, owing largely to Dainton and Ivin (22), the theory and consequences of reversible polymerizations have been worked out and the concept of a polymerization ceiling temperature has been defined (22). The significance of this concept is that for every concentration of monomer, there must be a temperature below which polymerization takes place and above which the reverse reaction is favored. And for every temperature there is an equilibrium monomer concentration that is thermodynamically determined. Thus the polymerization cannot proceed beyond a given conversion. We use the symbol T_c here to designate the temperature above which the equilibrium monomer concentration in the absence of solvent will be that of pure monomer; i. e. the temperature above which no polymerization will occur in the bulk.

It was once thought that the T_c of THF was very low, in fact near room temperature (23). However, in recent years, as catalyst systems have been improved and more intensive studies have been carried out, the presumed T_c has risen first to 60—70° (18, 24) and finally to 85 $\pm$ 2° C (25, 26, 27). The lower values were probably the result of working with systems where a true monomer-polymer equilibrium was not obtained. Possibly also, careful enough techniques were not used in the isolation of the lower molecular weight polymers obtained near the ceiling temperature. Precipitation in water cannot always be used because low molecular weight PTHF's are partially soluble in water.

The value 85 $\pm$ 2° C probably represents the true value of the T_c of THF. The same value has been obtained by several workers, albeit all using PF$_6^-$ gegenions. One group (25) derived it from equilibrium conversions obtained by both polymerization and depolymerization. The type of data used is shown in Table 2.

Table 2. *Equilibrium conversions by polymerization and depolymerization* (25)

Conditions			
Hours at T_1	Hours at T_2	Hours at T_1	% Conversion
24—25°	—	—	76.3
24—50°	24—25°	—	75.9
24—25°	24—50°	24—25°	75.8
48—50°	—	—	56.6
192—50°	—	—	55.8
24—25°	24—50°	—	56.9
24—50°	24—25°	24—50°	55.6
18—80°	—	—	13.4
18—50°	9—80°	—	12.6
24—25°	4—80°	48—25°	75.2
4—77°	—	—	17.9
72—25°	4—77°	—	18.2

When equilibrium conversions obtained in this manner are plotted against temperature a smooth curve (Fig. 1) which extrapolates to 84° C is obtained.

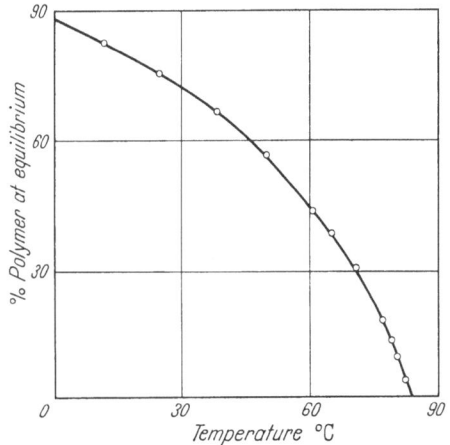

Fig. 1. Ceiling temperature in THF polymerization (25)

DAINTON and IVIN have related the equilibrium monomer concentration, M_e, to temperature in the equation:

$$\ln [M_e] = \frac{1}{T} \frac{\Delta H_p}{R} - \frac{\Delta S_p^0}{R} \qquad (1)$$

where ΔH_p is the heat of polymerization under the prevailing experimental conditions, and ΔS_p^0 is the entropy change for $[M_e] = 1$ mole liter^{-1}. It is seen that a plot of $\ln [M_e]$ versus $1/T$ should give a straight line whose slope is $\Delta H_p/R$ and whose intercept gives $-\Delta S_p^0/R$. The plot of $\ln [M_e]$ versus $1/T$ for THF is shown in Fig. 2. The equilibrium monomer concentrations corresponding to the amount of polymer isolated at various temperatures as discussed above were calculated assuming ideal solutions. From this plot ΔH_p was -4.58 kcal mole^{-1} and ΔS_p^0 was -17.7 cal deg^{-1} mole^{-1}. Values derived in a similar way have been reported by Sims (26), ΔH_p -4.3 ± 0.2 kcal mole^{-1}, ΔS_p^0 -17.0 ± 0.6 cal. deg.$^{-1}$ mole^{-1}, by Rozenberg et al. (24), ΔH_p -5.5 kcal mole^{-1}, ΔS_p^0 -20.8 cal. deg.$^{-1}$ mole^{-1}, and by Bawn et al. (18), ΔH_p $-5.3 \pm \pm 1.0$ kcal mole^{-1}.

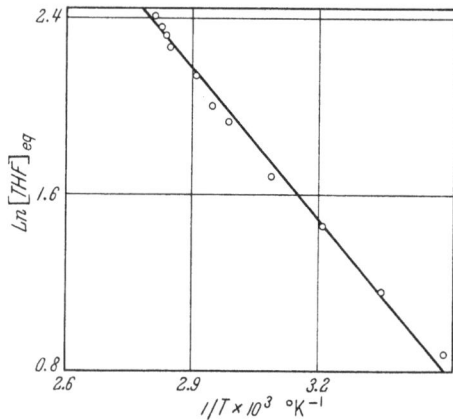

Fig. 2. Equilibrium monomer concentration as a function of temperature (25). Ln[THF]$_{eq}$. corresponds to ln[M_e] in equation (1)

Using the values of ΔH_p and T_c derived above and Dainton and Ivin's equation, $T_c = \Delta H_p / \Delta S_p$, we calculate an entropy of bulk polymerization $\Delta S_p = -12.8$ cal. deg.$^{-1}$ mole^{-1}. With this one can then get an estimate of ΔF_p^{298}; the value obtained is -800 cal. mole^{-1}, demonstrating that indeed THF is just able to polymerize at room temperature. With such a small negative free energy for THF, it is not surprising that substitution of even one hydrogen is enough to prevent polymerization.

Recently Leonard and Ivin (28) have pointed out that the application of equation (1) is valid only if the mixture of monomer and polymer behaves ideally over the range of compositions covered by the experiment. They reexamined some of the early data, making allowance for non-ideal mixing by use of the Flory-Huggins expression. They derived a equation for the free energy of polymerization in terms of the volume fractions of the polymer and monomer, φ_2 and φ_1 respectively, and χ, the polymer-monomer interaction parameter:

$$\Delta F_{lc} = RT \left[\ln \varphi_1 + 1 + \chi(\varphi_2 - \varphi_1)\right]. \tag{2}$$

However, they did not have the advantage of the above polymerization-depolymerization data. They had to adopt a rather complicated procedure to obtain a smooth curve for the variation of φ_2 with temperature. We have recalculated the data shown in Fig. 1 of fit the equation suggested by LEONARD and IVIN using the value $\chi = 0.3$, for which they

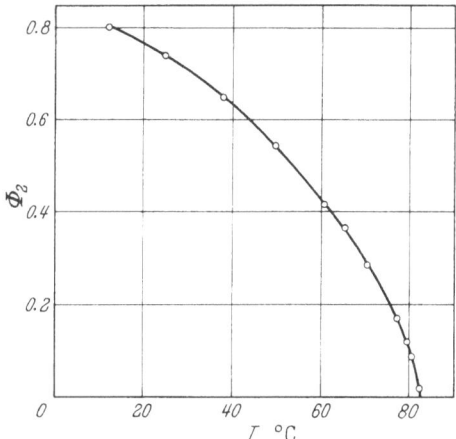

Fig. 3. Equilibrium volume fraction of polymer φ_2 in the bulk polymerization of tetrahydrofuran as a function of temperature

obtained the best fit. The results are seen in Figs. 3 and 4. Note that a smooth curve fits the experimental points (Fig. 3) and that a straight line is obtained when this curve is used to obtain $\Delta F/RT$ for the plot

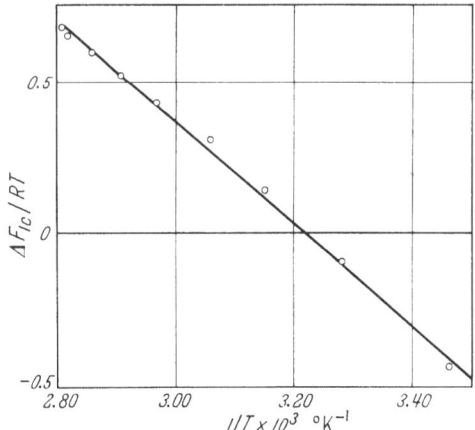

Fig. 4. Change of $\Delta F_{lc}/RT$ with temperature

of $\Delta F/RT$ versus $1/T$ (Fig. 4). ΔS_{lc} and ΔH_{lc} appear to be independent of temperature and can be obtained directly from the slope and intercept of the line. The values obtained are -3.3 kcal. mole^{-1} and -10.7 cal. deg.$^{-1}$

mole^{-1} for ΔH_{lc} and ΔS_{lc} respectively, in excellent agreement with those deduced by Leonard and Ivin (28), ΔH_{lc} -2.97 ± 1 kcal. mole^{-1}, ΔS_{lc} -9.75 ± 2 cal. deg.$^{-1}$ mole^{-1}.

But Leonard and Ivin point out, T_c is not given by $\Delta H_{lc}/\Delta S_{lc}$ since this takes no account of the partial molar free energies of the monomer and polymer. This can be seen in Fig. 4 which shows that $\Delta F_{lc}/RT$ is positive above 37° C, suggesting that polymerization will not occur above this temperature. Experimentally (Fig. 1) at 37° C approximately 65 % conversion to polymer is obtained and in fact polymerization continues to occur up to about 85° C. Practically, it would seem that either THF behaves ideally enough to justify the use of equation (1) for the derivation of meaningful heat and entropy terms for the derivation of T_c or non-idealities fortuitously cancel out. It should be understood that terms derived from equation 1 will include terms for the removal of monomer and addition of polymer to the polymer-monomer mixture.

III. Mechanism of polymerization

A. Propagation

It is now generally accepted (1, 25, 29) that the propagating species in the cationic polymerization of THF is the tertiary oxonium ion:

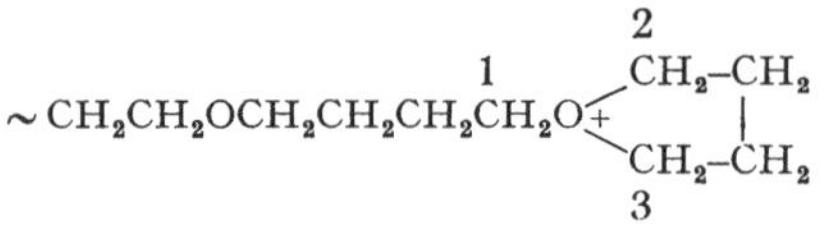

$$
\sim CH_2CH_2OCH_2CH_2CH_2O^+\!\!\!\begin{array}{c} \overset{2}{CH_2-CH_2} \\ | \\ CH_2-CH_2 \\ \overset{}{3} \end{array}
$$

Polymerization (propagation) occurs as a result of nucleophilic attack by monomer oxygen at a carbon alpha to the positive oxygen. Such attack at carbon (1), of course, results in no net change. But attack at either carbon 2 or carbon 3 results in growth; a new tertiary oxonium ion exactly like the original but one monomer unit longer is formed:

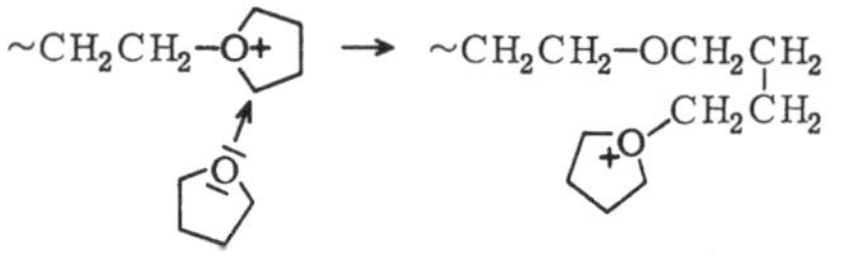

$$
\sim CH_2CH_2-\overset{+}{O} \longrightarrow \sim CH_2CH_2-OCH_2CH_2-CH_2CH_2 \tag{3}
$$

The depropagation reaction occurs by a similar nucleophilic attack by the penultimate oxygen atom (as shown by the dotted arrow) followed by the expulsion of monomer.

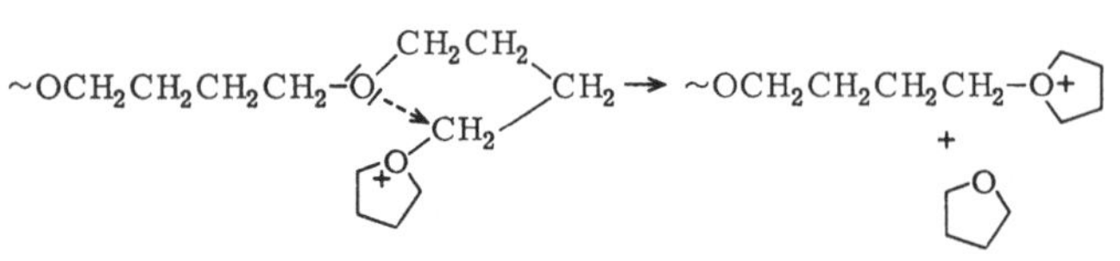

$$
\sim OCH_2CH_2CH_2CH_2-O \longrightarrow \sim OCH_2CH_2CH_2CH_2-\overset{+}{O} \tag{4}
$$

In some of the early work (3) it was suggested that the propagation proceeded by a preliminary opening of the ring to give a carbonium ion. Such a mechanism seems highly unlikely due to the improbable existence of carbonium ions when many oxygens are present. In their work with epoxides, PRICE and coworkers (30) have found that the cationic active center is *not* a carbonium ion but rather is always an oxonium ion. We expect that in the case of THF, oxonium ions would be even more strongly favored than with epoxides. In addition, ROZENBERG et al. point out (31), if carbonium ions form, it would be expected that the polymer chain would contain methyl branches. These would result from the isomerization of the unstable primary carbonium ion to the more stable tertiary or secondary carbonium ions. All studies to date indicate that PTHF is a linear polymer.

B. Initiation

Initiation requires the formation, in some manner, of a THF oxonium ion. The species R—O+ must be formed in order for the propagation reaction to take place. The different ways in which this species can be formed are conveniently grouped as follows:

1. Reaction with preformed trialkyl oxonium salts.
2. *In situ* formation of oxonium ion.
3. Addition of a carbonium ion.
4. Hydrogen abstraction.

1. Reaction with preformed trialkyl oxonium salts. The easiest mode of initiation to interpret involves transfer of an alkyl group from a preformed trialkyl oxonium ion salt to a THF molecule:

$$R-O_{R}^{+} \begin{matrix} R \\ \\ R \end{matrix} \quad X^{-} \quad + \quad O \begin{matrix} CH_2-CH_2 \\ | \\ CH_2-CH_2 \end{matrix} \quad \longrightarrow \quad R-O_{+}^{+} \begin{matrix} CH_2-CH_2 \\ | \\ CH_2-CH_2 \end{matrix} \quad X^{-} \quad + \quad O \begin{matrix} R \\ \\ R \end{matrix} \quad (5)$$

Like the propagation reaction, this probably occurs by a nucleophilic attack by the THF oxygen atom on a carbon alpha to the oxygen of the trialkyl oxonium ion salt. Commonly the triethyl-oxonium ion with a tetrafluoroborate gegenion has been used (3, 24, 32). MEERWEIN (3) has also used the hexachlorantimonate, tetrachloroferrate, and tetra-chloroaluminate gegenions.

The advantages of this type of initiator lie in the reasonably straight forward and unambiguous mode of initiation. Thus one can reasonably assume that each mole of initiator charged will initiate one mole of polymer chains. VOFSI and TOBOLSKY (32) neatly demonstrated that this is true and that the initation reaction is fast in a study using $Et_3O^+BF_4^-$ with C^{14} labeled ethyl groups. The disadvantages of this type of initiator lie in the relative difficulty of preparing the pure trialkyl oxonium salts, in the experimental problems of purifying and handling these moisture sensitive materials, and in the lability of the solutions of these salts (24).

2. In situ formation of oxonium ion. Because of the disadvantages of the trialkyl oxonium salts a number of workers have found it expedient to prepare the trialkyl oxonium ion or the THF oxonium ion directly in the polymerization charge. Instead of adding the preformed trialkyl oxonium salt, the reactants which will form the salt are charged.

The reaction used to prepare the trialkyl oxonium salts is that originally used by Meerwein (33). The reaction is carried out at dry ice temperatures. The intermediate product, which Meerwein calls an "inner oxonium ion salt" is not isolated. The preparation of $Et_3O^+BF_4^-$ is illustrated in equation 6.

$$
\begin{array}{c}
CH_3-CH_2 \\
\quad\quad O\cdot BF_3 \\
CH_3-CH_2
\end{array}
+
\begin{array}{c}
CH_2-CH-CH_2Cl \\
\diagdown\ \diagup \\
O
\end{array}
\longrightarrow
\begin{array}{c}
CH_2-CH_2 \quad\quad CH_2-CH^{CH_2Cl} \\
\quad\quad {}^+O \quad\quad O \\
CH_3-CH_2 \quad BF_3^-
\end{array}
$$

$$
\begin{array}{c}
CH_3CH_2 \quad\quad CH_2CH_3 \\
\quad\quad {}^+O \\
CH_3CH_2 \quad BF_4^-
\end{array}
+
\begin{array}{c}
CH_3CH_2 \quad CH_2-CH^{CH_2Cl} \\
\diagdown\ \diagup \quad | \\
O \quad O \\
\quad\quad | \\
\quad\quad BF_2
\end{array}
\longleftarrow
\begin{array}{c}
CH_3CH_2 \\
\quad\quad O\cdot BF_3 \\
CH_3CH_2
\end{array}
\quad (6)
$$

$$
3CH_3CH_2OCH_2\overset{CH_2Cl}{CH}OBF_2 \longrightarrow (CH_3CH_2OCH_2\overset{CH_2Cl}{CH}-O)_3B + 2BF_3 \quad (6a)
$$

a) *In the presence of added promotor.* Frequently in the *in situ* formation of oxonium ion BF_3 or BF_3 etherate is used as initiator in conjunction with epichlorohydrin (ECH) or other epoxides such as ethylene oxide or propylene oxide [Ofstead (34), Rozenberg et al. (24, 31), Saegusa et al. (35)]. Saegusa and Furukawa (1) call the Lewis acid component "the catalyst", and the epoxide additives "promotors". In addition to the epoxides Saegusa et al. (35) have shown that diketene, β-propiolactone, and 3,3-bis-chloromethyloxetane are effective. Promotors have been used in other Lewis acid catalyst systems also. Miller (11) found that SiF_4 is a catalyst, but only when used in conjunction with ethylene oxide. Sims (37) found that ECH was very helpful in PF_5 initiation of THF polymerizations.

Since THF is a much stronger base than ether, (36, 38) it doesn't matter whether gaseous BF_3 is used or if it is charged in the form of BF_3 etherate. It will very rapidly be present in a bulk THF system in the form of the $BF_3 \cdot THF$ complex. It is this complex, then, which must interact with the promotor to effect initiation. The reaction of this complex with the promotor may be analogous to equation 6 above, proceeding as follows:

$$
\begin{array}{c}
CH_2-CH_2 \\
\vert \quad \diagdown O \cdot BF_3 \\
CH_2-CH_2 \diagup
\end{array}
+
\begin{array}{c}
CH_2-CH-CH_2Cl \\
\diagdown O \diagup
\end{array}
\longrightarrow
\begin{array}{c}
\qquad\qquad CH_2Cl \\
\qquad\qquad \diagup \\
CH_2-CH_2 \quad CH_2-CH \\
\vert \quad \diagdown \!+\!O \diagup \diagdown O \diagup \\
CH_2-CH_2 \diagup \quad BF_3^-
\end{array}
$$

$$
\begin{array}{c}
CH_2CH_2 \quad CH_2CH_2CH_2CH_2 \quad CH_2-CH \\
\vert \quad \diagdown \!+\!O \diagup BF_4^- \qquad\qquad \diagdown O \diagup \diagdown O \diagup \\
CH_2CH_2 \diagup \qquad\qquad\qquad BF_2
\end{array}
\longleftarrow
\begin{array}{c}
\qquad\qquad CH_2Cl \\
\qquad\qquad \diagup \\
CH_2-CH \\
\diagdown O \diagup \diagdown O \diagup \\
BF_2
\end{array}
\Bigg\vert
\begin{array}{c}
CH_2CH_2 \\
\vert \quad \diagdown O \cdot BF_3 \\
CH_2CH_2 \diagup
\end{array}
\quad (7)
$$

The only difference from equation 6 is that the two ethyl groups are now joined and only one species is formed in the final step. This predicts an endgroup derived from ECH, and produces the required tertiary oxonium ion associated with a simple gegenion. On considering this type of initiation some authors (3, 16, 34) stop at the first intermediate and say that this inner oxonium salt is then capable of propagating in the usual manner. We do not favor this interpretation because it requires association either between the two ends of each polymer molecule or between two polymer chains. The latter should lead to an association and the formation of a network system during the polymerization. This is not consistent with observations. The path suggested in equation 7 has an excellent analogy in the preparation of the trialkyl oxonium salt (equation 6) and it does lead to the more satisfying situation of having the growing oxonium ion associated with a small unencumbered gegenion.

Moreover, the evidence all seems to suggest that although an inner oxonium salt is probably a true intermediate, it is a very reactive species and soon reacts to form the trialkyl oxonium salt. MEERWEIN (33) in some cases was able to isolate and characterize this intermediate. But in the case of the product from boron trifluoride etherate and ECH it decomposed before it could be warmed to room temperature. Furthermore, ROZENBERG (24) reports that the reaction of boron trifluoride etherate with ECH is practically instantaneous, with a quantitative yield of $Et_3O^+BF_4^-$. The formation of $Et_3O^+BF_4^-$ by the direct combination of $Et_2O \cdot BF_3$ with ethyl fluoride is a very slow reaction (33). It would appear that the use of ECH provides, in forming the inner oxonium salt, a low energy pathway for this reaction.

SIMS (37) made a similar interpretation of the results of his study of ECH promoted PF_5 catalyzed THF polymerizations. The scheme he proposed was:

$$
\begin{array}{c}
\text{CH}_2\text{CH}_2 \\
|\quad\quad\quad\diagdown \\
\quad\quad\quad\quad \text{O} \cdot \text{PF}_5 \\
|\quad\quad\quad\diagup \\
\text{CH}_2\text{CH}_2
\end{array}
+
\begin{array}{c}
\text{CH}_2\text{-CH-CH}_2\text{Cl} \\
\diagdown\quad\diagup \\
\text{O}
\end{array}
\longrightarrow
\begin{array}{c}
\text{CH}_2\text{-CH}_2 \quad\quad \text{CH}_2 \\
|\quad\quad\quad {}^+\text{O}\cdots\diagdown \\
\quad\quad\quad\quad\quad\quad \text{CH-CH}_2\text{Cl} \\
\text{CH}_2\text{-CH}_2 \quad \text{PF}_5\text{O}
\end{array}
$$

$$
\begin{array}{c}
\text{CH}_2\text{CH}_2 \\
|\quad\quad\quad\diagdown \\
\quad\quad\quad\quad \text{O} \cdot \text{PF}_5 \\
|\quad\quad\quad\diagup \\
\text{CH}_2\text{CH}_2
\end{array}
\tag{8}
$$

$$
\begin{array}{c}
\text{CH}_2\text{CH}_2 \\
|\quad\quad\quad {}^+\text{OCH}_2\text{CH(CH}_2\text{Cl)OPF}_4 \\
|\quad \\
\text{CH}_2\text{CH}_2 \\
\quad\text{PF}_6^-
\end{array}
\xrightarrow{\text{THF}\cdot\text{PF}_5}
\begin{array}{c}
\text{CH}_2\text{CH}_2 \\
|\quad\quad\quad {}^+\text{OCH}_2\text{CH(CH}_2\text{Cl)OPF}_3 \\
|\quad \\
\text{CH}_2\text{CH}_2 \\
\quad\text{PF}_6^-
\end{array}
\quad\text{O}{}^+
\begin{array}{c}
\text{CH}_2\text{CH}_2 \\
\diagdown \\
|\quad \\
\diagup \\
\text{CH}_2\text{CH}_2 \\
\text{PF}_6^-
\end{array}
$$

Note that the first two steps are essentially the same as equation 7. Sims continues and writes the third step in order to explain his observed ratio of number of chains formed to number of ECH and PF_5 molecules charged. Sims showed by chlorine analysis of the polymers that for ECH/PF_5 ratios of 0 to 1/4 all of the ECH was immediately incorporated into the polymer. At higher ratios of ECH/PF_5, the rate of incorporation of ECH was much slower, suggesting that the ECH was no longer acting as a promotor and was only copolymerizing. (Cf. also kinetics Section IV.) Note that polymerization of the inner oxonium salt itself would require an ECH/PF_5 ratio of 1. Thus Sims' data clearly suggest that the propagating species is not the inner oxonium salt.

In the past few years the use of aluminum alkyls as catalysts for cyclic ether polymerizations has received much attention. Two different mechanisms have been proposed to explain the catalytic activity of the aluminum alkyl catalysts. Saegusa, Imai, and Furukawa (15) suggest that a cationic mechanism is produced. They feel it is not related to the coordinate anionic mechanism presumed to take place with related catalyst systems used for aldehydes and epoxides. They propose that the Lewis acid first reacts with adventitious water to form a Brönsted acid.:

$$
\text{Et}_3\text{Al} + \text{H}_2\text{O} \longrightarrow \text{H}^+\overline{\text{Al}}\text{Et}_3\text{OH}
$$

$$
\overset{\text{ECH}}{\swarrow} \quad \overset{\text{THF}}{\searrow}
\tag{9}
$$

$$
\begin{array}{c}
\text{ClCH}_2\text{-CH} \\
|\quad\quad\diagdown {}^+ \\
\quad\quad\quad\text{OH} \quad \overline{\text{Al}}\text{Et}_3\text{OH} \\
|\quad\quad\diagup \\
\text{CH}_2
\end{array}
\quad\quad
\begin{array}{c}
\text{CH}_2\text{-CH}_2 \\
|\quad\quad\quad\diagdown {}^+ \\
\quad\quad\quad\quad\text{OH} \quad \overline{\text{Al}}\text{Et}_3\text{OH} \\
|\quad\quad\quad\diagup \\
\text{CH}_2\text{-CH}_2
\end{array}
$$

The Brönsted acid can then react with both the ECH promotor and the THF monomer to form the dialkyl oxonium ions shown. Either of these can react with THF to produce the propagating trialkyl oxonium species. But Saegusa et al., argue that the ECH species will undergo this reaction much more readily than the dialkyl oxonium ion derived from THF. Thus an acceleration is observed when ECH is used.

This group also reports the activating effect of water on these aluminum alkyl catalysts. They call this water a "modifier" because they feel its function is to form new modified Lewis acids from the aluminum alkyls. They then consider that these complex modified Lewis acids react in a scheme just like the one outlined above (equation 9).

In contrast, WEISSERMEL and NÖLKEN (16) suggest that the ECH reacts directly with the alkyl aluminum and THF to form an inner oxonium salt:

$$\underset{\underset{CH_2}{|}}{ClCH_2-CH} \Big\rangle O \ + \ AlR_3 \ \rightarrow \ \left[\underset{\underset{CH_2\ +}{|}}{\overset{ClCH_2}{\underset{}{\diagdown}}} \underset{}{\overset{}{CH-O\bar{A}lR_3}} \right] \xrightarrow{\ THF\ } \underset{\underset{CH_2Cl}{|}}{\overset{\bar{O}AlR_3}{\diagup}} \underset{}{\overset{}{+O-CH_2-CH}} \qquad (10)$$

They suggest that it is this ion which propagates to form polymer.

SAEGUSA et al. (66) have recently made a detailed study of the behavior of promotors in the AlEt$_3$–H$_2$O-promotor system. They shortstopped a polymerization with sodium methoxide after two minutes of reaction. They detected the presence of HOCH(CH$_2$Cl)CH$_2$O(CH$_2$)$_4$OCH$_3$ by vapor phase chromotographic analysis. Hence, they now propose an initiation scheme similar to WEISSERMEL and NÖLKEN's. SAEGUSA et al. state that isolation of this alcohol indicates the presence of the product of equation 10 in the early polymerization mixture.

However, we suggest carrying the scheme one step further in a manner analogous to equations 7 and 8:

$$\underset{}{\overset{OAlR_3}{\diagup}} \underset{}{\overset{}{+O-CH_2-CH}} \underset{CH_2Cl}{\diagdown} \ + \ \underset{}{\overset{}{O \cdot AlR_3}} \ \rightarrow \ \underset{}{\overset{OAlR_2}{\diagup}} \underset{}{\overset{}{+O-CH_2CH}} \underset{CH_2Cl}{\diagdown} \qquad (11)$$
$$AlR_4^-$$

It should be noted that the alcohol isolated by SAEGUSA (66) is also entirely consistent with equation 11. This equation predicts that the gegenion in an AlR$_3$/ECH catalyzed polymerization is AlR$_4^-$. This is not an unreasonable species. It is strictly analogous to the AlCl$_4^-$ ion used by MEERWEIN (3). Furthermore, in a review of alkyl aluminum compounds, ZIEGLER (39) discusses salts such as NaAlEt$_4$, NaAlEt$_3$OR, KAlEt$_2$Cl$_2$, KAlEtCl$_3$, Na(AlEt$_3$OAlEt$_2$), further indicating the plausibility of such complex aluminum anions.

One final class of initiators seems to belong in this group. The use of the Lewis acid/orthoester combination reported by MEERWEIN (3) is similar to the Lewis acid/promotor systems just discussed, in that the tertiary oxonium ion is generated *in situ* from the reactants. In this case, however, one first needs to consider the preparation of the "tertiary carboxonium salts" (3, 10). These salts are formed by interaction of orthoesters with Lewis acids:

$$3RC(OR')_3 + 4BF_3 \rightarrow 3 \left(RC \begin{array}{c} OR' \\ OR' \end{array} \right)^+ BF_4^- + B(OR')_3 \qquad (12)$$

This type of resonance stabilized oxonium ion is the chain carrier proposed by Dreyfuss and Dreyfuss (25) in the transfer reaction with trimethylorthoformate (Section III D).

In the use of the BF_3/orthoester combination as a catalyst for the polymerization of THF, Meerwein (3) considers that this reaction proceeds only part way:

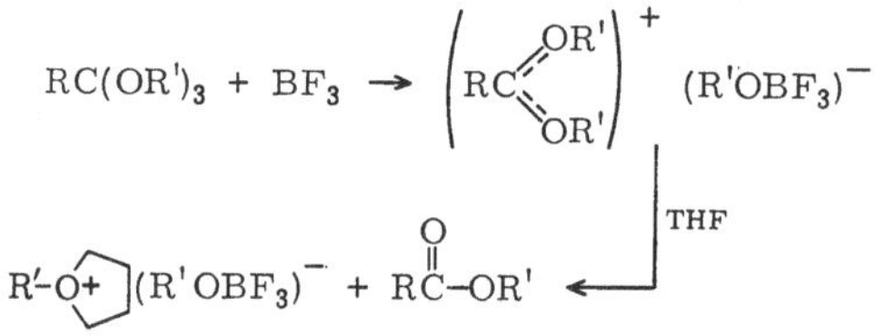

$$RC(OR')_3 + BF_3 \rightarrow \left(RC \begin{array}{c} OR' \\ OR' \end{array} \right)^+ (R'OBF_3)^-$$

$$\Big\downarrow \text{THF}$$

$$R'-\overset{+}{O}\Big](R'OBF_3)^- + R\overset{O}{\overset{\|}{C}}-OR' \longleftarrow \qquad (13)$$

However, perhaps even in this case the reaction takes the course of equation 12 rather than equation 13. The gegenion would then again be BF_4^- rather than the $R'OBF_3^-$ suggested by Meerwein.

b) In the absence of added promotor. A similar mechanism can be advanced for the polymerization of THF by the Friedel-Crafts halides alone. It is not unreasonable to suppose that in the case of a very reactive Lewis acid like PF_5, THF itself can function as the "promotor" molecule, serving the function, albeit at a slower rate, of the ECH. If this is the mode of initiation, the reactions would be:

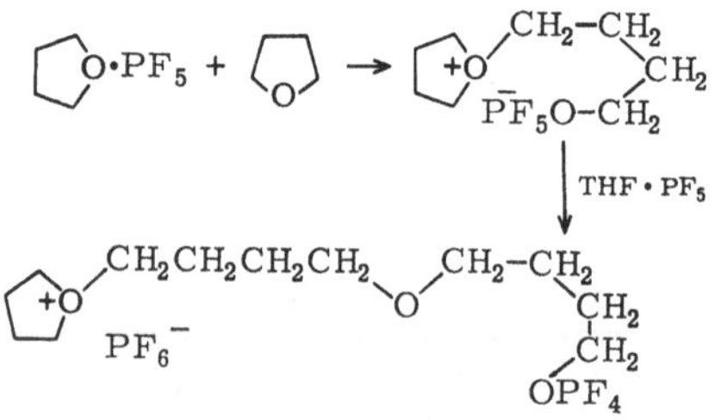

$$\begin{array}{c} \overset{O \cdot PF_5}{\bigcirc} + \overset{}{\bigcirc_O} \rightarrow \overset{+O}{\bigcirc} \overset{CH_2-CH_2}{\underset{PF_5O-CH_2}{\diagdown}} CH_2 \\ \Big\downarrow \text{THF} \cdot PF_5 \\ \overset{+O}{\bigcirc} \overset{CH_2CH_2CH_2CH_2}{\underset{PF_6^-}{\diagdown}} O \overset{CH_2-CH_2}{\underset{OPF_4}{\diagdown}} \overset{CH_2}{\underset{CH_2}{|}} \end{array} \qquad (14)$$

In the only study to date of the mechanism of PF_5 initiation, Sims (40) could conclude only that polymerization of THF with PF_5 as catalyst is a very complex system. He found that there appears to be a slow initiation reaction which does not reach completion until at least half of the polymerization is over. Sims also studied the effect of water on the polymerization, but could not fully explain the complex results. He suggested that two competing effects of water, co-catalysis and destruction of catalyst, might be involved.

Similar equations can of course be written for initiation by SbCl$_5$ alone. MEERWEIN (3) who first reported its use, considers two possibilities for initiation. One is the formation of an "auto complex", $\left[\bigcirc\!\!\!\!\!\text{O--SbCl}_4 \right]^+$ SbCl$_6^-$ and the other is the formation of an inner oxonium complex, Cl$_5\overline{\text{SbO}}$(CH$_2$)$_4$-$\overset{+}{\text{O}}$. MEERWEIN appears to favor the formation and polymerization of the inner oxonium complex. ROZENBERG and co-workers (41), have recently reexamined this question and conclude that THF initiation in this case can be represented as follows:

$$2 \ \text{SbCl}_5 \cdot \text{O} \bigcirc \longrightarrow \text{SbCl}_4\text{O}(\text{CH}_2)_4\text{-}\overset{+}{\text{O}} \bigg] \ \text{SbCl}_6^- \qquad (15)$$

and present kinetic evidence that initiation seems to involve the formation of an "auto-complex". Note that really the two possibilities suggested by MEERWEIN and ROZENBERG may be the same. Equation 14 provides a path whereby the "auto-complex" is formed *via* the inner oxonium complex.

As SIMS found in the PF$_5$ case ROZENBERG (41) notes that initiation of active chains from SbCl$_5 \cdot$ THF is a slow process. Even at the end of the polymerization reaction only a small fraction of initiator has been consumed.

It is interesting to consider the BF$_3$ case. BF$_3$ in the absence of promotor molecules does not initiate polymerization of THF unless very high concentrations of BF$_3$ are employed [BURROWS (42)]. Perhaps the equilibrium constant of the reaction:

$$\bigcirc\!\!\!\!\text{O} \cdot \text{BF}_3 \ + \ \bigcirc\!\!\!\!\text{O} \ \rightleftharpoons \ \bigcirc\!\!\!\!\overset{+}{\text{O}} \overset{\text{CH}_2\text{-CH}_2}{\underset{\text{BF}_3\text{O-CH}_2}{\diagdown\!\!\!\diagup\text{CH}_2}} \qquad (16)$$

is so low that the concentration of the products do not become significant until a BF$_3$ concentration in excess of 5 mole percent is reached.

There is still much work to be done in clearly defining the mechanism of initiation and the gegenions formed in the *in situ* preparation of trialkyl oxonium salts. Clearly, however, for theoretical studies of the polymerization the use of preformed trialkyl oxonium salts is to be preferred. For the preparation of polymer, on the other hand, the more convenient *in situ* method may be preferred, particularly if the desired product is a polytetramethylene glycol.

3. Addition of a carbonium ion. The scheme:

$$\text{R}^+\text{X}^- \ + \ \text{O}\bigcirc \ \longrightarrow \ \text{R-}\underset{\text{X}^-}{\overset{+}{\text{O}}}\bigcirc \qquad (17)$$

has not proved to be very practical, for it is difficult to obtain stable species of the $R^+ X^-$ type. The most stable species of this type studied, the triphenylmethyl salts, do not initiate by this mechanism (see 4. below). The only examples that seem to remain are combination of an acid halide with a Lewis acid. MEERWEIN (3) reports a number of combinations of this type that are effective catalysts for THF polymerization. A representative example is the combination CH_3COCl and $SbCl_5$ which is presumed to produce $CH_3CO^+SbCl_6^-$. Initiation is then pictured as (3, 41):

$$CH_3CO^+SbCl_6^- \ + \ O\bigcirc \ \rightarrow \ CH_3\overset{O}{\underset{SbCl_6^-}{\overset{\|}{C}}}-O{+}\bigcirc \qquad (18)$$

In a recent study of this system ROZENBERG (41) found that the rate of reaction of the dialkyl acyl oxonium ion with THF is less than the rate of propagation, probably as a result of conjugation of the oxonium ion with the carbonyl group.

4. Hydrogen abstraction. Initiation by $Ph_3C^+SbCl_6^-$ (17, 18) and more recently $Ph_3C^+PF_6^-$ (20) has been the subject of a number of investigations. During the early stages of these investigations (17, 19) it was reasonably considered possible that initiation occurred by addition of the carbonium ion to the oxygen of THF to generate the oxonium ion $Ph_3C\overset{}{O}{+}\bigcirc$. Subsequent studies by KUNTZ (43, 44) showed that Ph_3C- was not an endgroup of the polymer. KUNTZ (43) demonstrated that $Ph_3C^+SbCl_6^-$ and $(Ph_2\overset{+}{C}C_6H_4CH_{\bar{2}})_{\bar{2}}(SbCl_6^-)_2$ when used in equivalent amounts of carbonium ion generated the same molecular weight PTHF; two growing ends were not joined by the initiator molecule in the latter case. More recently (44), he demonstrated the formation of Ph_3CH directly by using NMR. The alternative mechanism (19) involving a hydride ion abstraction from the THF molecule was indicated (43).

BAWN et al. (19), and KUNTZ (43) suggested that after hydride ion abstraction the resulting species reacts with another molecule of THF to form a growing polymer molecule which has an acetal end group:

$$\bigcirc_{O}{+} \ + \ O\bigcirc \ \rightarrow \ \bigcirc_O{-}O{+}\bigcirc \qquad (19)$$

Initiation by $p\text{-}ClC_6H_4N_2^+PF_6^-$ also seemed to occur via a hydride ion abstraction. DREYFUSS and DREYFUSS (25) showed that the expected product of hydride ion abstraction, chlorobenzene, is formed in the decomposition of $p\text{-}ClC_6H_4N_2^+PF_6^-$ in 2-MeTHF and the product of thermal decomposition, p-chlorofluorobenzene, was absent. In this case abstraction involving a six membered cyclic intermediate was suggested:

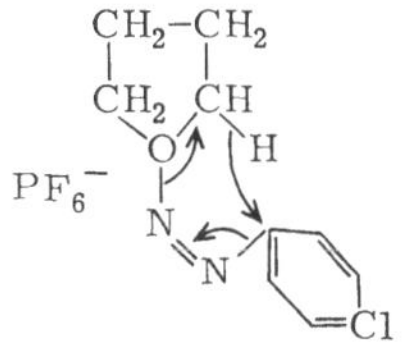

$$
\begin{array}{c}
CH_2-CH_2 \\
| \quad\quad | \\
CH_2 \quad CH
\end{array}
$$

PF$_6^-$

Recent NMR studies by DREYFUSS, WESTFAHL, and DREYFUSS (45) show that the interaction between Ph$_3$C$^+$SbCl$_6^-$ and 2-MeTHF or THF appears to result in the formation of HSbCl$_6$ in addition to Ph$_3$CH. Evidence for the formation of furans was also reported. They suggest that the reaction is a dehydrogenation rather than a hydride ion abstraction. On the basis of stoichiometry they further suggested that, in the case of THF, dehydrogenation may yield dihydrofuran which is then rapidly removed from the system by acid catalyzed polymerization. The initiation scheme that they propose for Ph$_3$C$^+$SbCl$_6^-$ is as follows:

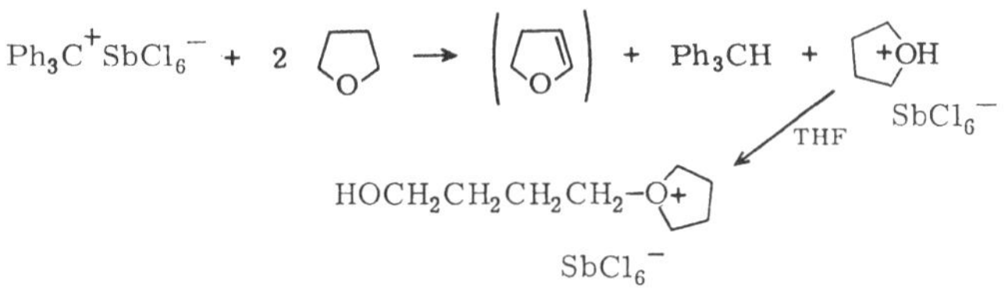

$$
Ph_3C^+SbCl_6^- + 2 \quad \longrightarrow \quad + Ph_3CH +
$$

$$
HOCH_2CH_2CH_2CH_2-O^+ \qquad\qquad (20)
$$

SbCl$_6^-$

The implication is that where hydride ion abstraction is indicated, the probable true initiator is the acid HSbCl$_6$ or HPF$_6$. DREYFUSS et al. further point out that one cannot make the apriori assumption that the reactivity of the dialkyl oxonium ion (THF · HSbCl$_6$) is comparable to the reactivity of the trialkyl oxonium ion (i.e. the propagating species). In fact, it may be slower (46). Thus any assumptions about the rate of initiation, or about the number of active centers formed, or any attempts to correlate the degree of polymerization of the resulting polymer with the amount of carbonium ion salt or aryl diazonium salt initiators charged should be made with extreme caution. Again, for theoretical studies of the polymerization, the use of preformed trialkyloxonium salts is to be preferred.

C. Termination

There have been numerous claims (16, 18, 21, 26, 31, 48, 49) that the polymerization of THF in the presence of certain catalysts proceeds without termination. A number of authors hasten to add that termination and transfer are not important under the conditions of their experiments but that slow termination or transfer reactions may exist. Several kinds of termination reactions that can occur are discussed in this section and transfer reactions are considered in the following section.

1. With impurities. As in all ionic polymerizations, a high level of purity of the monomer and other components of the polymerization system needs to be achieved before THF polymerization can occur. Water, acid, oxygen, amines and the like cause termination by reaction with the cation complex and in fact are used to terminate the reaction. However, with some systems, notably those containing the PF_6^- gegenion, the oxonium ion-metal ion complex appears to be quite stable and polymerizations can be very successfully carried out under nitrogen, although high vacuum work is still recommended for theoretical studies. Studies in solvent are complicated in part because of possible reaction with the solvent and also because of the lowering of the ceiling temperature followed by the subsequent need to use ever lower temperatures where the propagation rate is slow.

2. With an anionic "living" polymer. Recently Berger, Levy, and Vofsi (*49*) have reported the preparation of block copolymers by mutual termination of anionic and cationic "living" systems. They have demonstrated this interesting method by terminating "living" anionic polystyrene with "living" PTHF:

$$R_1\text{-\!\!\wedge\!\!\wedge\!\!-}CH_2\text{-}\underset{\underset{\bigcirc}{|}}{C}H^-\,Na^+ + BF_4^-\,\left[\overset{+}{\underset{O}{}}\text{-\!\!\wedge\!\!\wedge\!\!-}R_2\right. \longrightarrow$$

$$R_1\text{-\!\!\wedge\!\!\wedge\!\!-}CH_2\text{-}\underset{\underset{\bigcirc}{|}}{C}H\text{-}CH_2CH_2CH_2CH_2\text{-}O\text{-\!\!\wedge\!\!\wedge\!\!-}R_2 + NaBF_4 \tag{21}$$

The block copolymer was a transparent rubbery substance with properties quite different from those of the homopolymers. The success of this reaction and the possibility of expanding it to other pairs of cationic and anionic monomers, provides a possible way to a variety of new "tailor-made" polymers.

3. With gegenion. The gegenion associated with the propagating oxonium ion has a great effect on the course of the polymerization of THF. It determines to a large extent the final degree of conversion at a given temperature and the molecular weight that will be reached. Therefore, we need to consider carefully the properties of the oxonium ion-gegenion complex. Simple anions like Cl^- are not suitable gegenions because the trialkyl oxonium ion complex, $R_3O^+X^-$, is unstable with respect to the ether and alkyl halide, R_2O and RX (*53*). Thus in order to obtain a stable species of this type the complex ions have been applied.

a) Tetrafluoroborate, BF_4^-. There are some indications in the literature which suggest that at $0°$ C, BF_4^- is a suitably stable gegenion. Vofsi and Tobolksky (32) found that with a monomer concentration in the range 4 to 10 mole/l and a catalyst concentration below 3.0×10^{-2} mole/l, they could interpret their kinetic data without including a termination reaction. Outside that range they felt a termination term would be required. Ofstead (34) showed that the initial polymers obtained at 0° C with BF_4^- gegenions had a very narrow molecular weight distribution.

But at room temperature and above the BF_4^- gegeion is not so stable. Meerwein (3) has proposed that it is possible for the BF_4^- gegenion to react with the growing oxonium ion and lead to termination in the following way:

$$\text{~CH}_2\text{CH}_2\text{-}\overset{+}{\underset{BF_4^-}{\text{O}}} \longrightarrow \text{~CH}_2\text{CH}_2\text{-O}\overset{CH_2-CH_2}{\underset{CH_2F}{\diagup}}\underset{CH_2}{\diagdown} + BF_3 \qquad (22)$$

The possibility of this reaction probably accounts for the fact that at room temperature and above the yields reported in the bulk system using BF_3 based catalysts all have led to a low ceiling temperature. It should be noted here that Rozenberg et al. (31) do not believe that the termination reaction is important with the BF_4^- ion. This is hard to reconcile with the fact that at 25° C and above their ultimate conversions are 5—10% lower than those obtained with the PF_6^- gegenion.

b) *Hexachloroantimonate*, $SbCl_6^-$. The $SbCl_6^-$ gegenion has only been studied extensively at room temperature and above. At 20—25° C, the ultimate conversion that is obtained (18, 41) is the equilibrium conversion. However, at higher temperatures, the ultimate conversions again fall below the equilibrium conversions (18, 20), suggesting that some termination occurs.

c) *Hexafluorophosphate*, PF_6^-. Experiments near room temperature with PF_5 (26) and 4-ClPhN$_2$PF$_6^-$ (25) indicate that at this temperature THF polymerizations proceed without termination. Using the PF_5 catalyst, Sims (40) was able to show that polymerization continued at essentially the same rate after second additions of monomer. Fluorine content of the polymers he prepared were found to range from zero to a few parts per million. Dreyfuss and Dreyfuss (25) were able to carry out polymerization-depolymerization-repolymerization experiments in the presence of PF_6^- gegenions (Table 2, Section II). They (21) were also able to continue polymerization on the addition of either THF or other cyclic ethers and found that molecular weight increased directly as catalyst concentration decreased. The equilibrium conversions obtained by Sims and by Dreyfuss and Dreyfuss are in good agreement over the whole range. All these results combined illustrate that in the presence of PF_6^- gegenions there is little or no termination in THF polymerizations under most conditions.

There are some indications that at 70—80° C there may be some termination occurring even with the PF_6^- gegenion. In polymers produced at 70° C. Sims (26) was able to detect some fluorine which he suggests may have formed from a reaction of the type given below:

$$\sim (CH_2)_4-O-(CH_2)_4 \sim + PF_6^- \rightarrow \sim (CH_2)_4-OPF_5^- + \sim (CH_2)_4F \quad (23)$$

Another possibility would be the reaction analogous to equation (22):

$$\underset{PF_6^-}{\sim\sim CH_2CH_2-O^+} \rightarrow \sim\sim CH_2CH_2-O-(CH_2)_4F + PF_5 \quad (24)$$

As we have stated before (Section III B 4) it is probable that $Ph_3C^+PF_6^-$ and $4\text{-ClPhN}_2^+PF_6^-$ initiate polymerization by the same mechanism and lead to the same propagation reaction. These suggestions are reinforced by the data (20, 52) shown in Table 3. For similar catalyst concentrations, polymers of similar intrinsic viscosity are obtained.

Table 3. *Comparison of initiators*

Catalyst	10^3 [cat] mole/l	Temperature °C	$[\eta]$dl/g★
$Ph_3C^+PF_6^-$	5.41	25	7.0
$4\text{-ClPhN}_2^+PF_6^-$	5.57	25	7.5
$Ph_3C^+PF_6^-$	3.94	25	9.1
$4\text{-ClPhN}_2^+PF_6^-$	4.0	25	9.4
$Ph_3C^+PF_6^-$	5.02	50	5.8
$4\text{-ClPhN}_2^+PF_6^-$	5.5	50	4.9—5.6
$Ph_3C^+SbCl_6^-$	4.93	25	1.76
$Ph_3C^+SbCl_6^-$	3.53	25	2.06
$Ph_3C^+SbCl_6^-$	4.0	50	1.54

★ Benzene at 25° C.

$Ph_3C^+SbCl_6^-$ probably initiates by the same mechanism as the PF_6^- salts. But in this case much lower intrinsic viscosity polymers are obtained from similar catalyst concentrations (Table 3) (20). Apparently more termination and transfer reactions occur during propagation with $SbCl_6^-$ gegenions compared to propagations with PF_6^- gegenions (20). Comparable data for the BF_4^- gegenion is not available.

d) Relative stabilities of the gegenions. Since we are dealing with the same propagating ion in all the cases of THF polymerizations, the different effects noticed with different catalysts cannot be a function of the oxonium ion and must be related in some way to the characteristics of the associated gegenions themselves. From the polymerization results just discussed one would predict the following order of stabilities:

$$PF_6^- > BF_4^- \geqq SbCl_6^- \quad (25)$$

"Stability" as used in this sense refers to the ability of the gegenion to avoid chemical reaction with the oxonium ion itself or with other components in the polymerization mixture such as traces of water or solvent.

Though the literature on this problem is surprisingly limited and some direct studies would be desirable, there are a few studies which indicate that the above conclusions are reasonable. The factors which need to be considered in determining stability include charge/radius ratio, polarizability, the ability to use empty d orbitals for back bonding, and lattice energy (54).

aa) Electronic structure. In order to orient ourselves, it might be helpful to consider the location in the periodic table and the electronic configuration of the elements under consideration (Tables 4 and 5).

Table 4. *Location in periodic table*

Group III	Group V		Group VII
		Group V b	
B	N		F
Al	P		Cl
		As	
		Sb	

Table 5. *Electronic configuration of atoms in normal state*

Element	At. No.	Covalent Radii $(55, 56)$	Configuration
B	5	0.88	$1s^2\,2s^2\,2p^1$
P	15	1.10 (1.11)	$1s^2\,2s^2\,2p^6\,3s^2\,3p^3$
Sb	51	1.36 (1.41—1.45)	$1s^2\ldots4s^2\,4p^6\,4d^{10}\,5s^2\,5p^3$
F	9	0.64 (0.72)	$1s^2\,2s^2\,2p^5$
Cl	17	0.99 (1.00)	$1s^2\,2s^2\,2p^6\,3s^2\,3p^5$

It has been postulated (56) that the presence of d orbitals in the metal leads to high ionic character in the bonds and stronger bonding of the halogen atom and greater stability. From the structure we see that B will have a valence of $+3$ when fully substituted. B can accomodate 4 fluorines as in BF_4^- and the complex itself will be charged -1. Excitation of the valence electrons will require fairly high energies. P and Sb, on the other hand, already have their d orbitals in use. Both are in group V and have a valence of $+5$ when fully substituted, higher than B in BF_4^-. Each can accomodate 6 halogens and when so substituted, the resulting complex has a charge of -1, exactly the same as BF_4^-. Thus, from a consideration of electronic structure alone, one might expect PF_6^- and $SbCl_6^-$ to be more stable than BF_4^-.

However, it is also necessary to consider the Cl and F. Fluorine is the most electronegative atom known and as a result its compounds are very stable. Thus one might predict that PF_6^- and BF_4^- would be more stable than $SbCl_6^-$.

bb) Ionic structure. All of the ions under consideration have symmetrical structures, the BF_4^- being tetrahedral (*57*) and the other two having a hexagonal arrangement of the halogens around the metal ion (*56, 58*) (Fig. 5). All of these symmetrical ions possess relatively great stability, form strong acids, and are difficult to deform. On the basis of ionic structure alone it is difficult to make any predictions about stability.

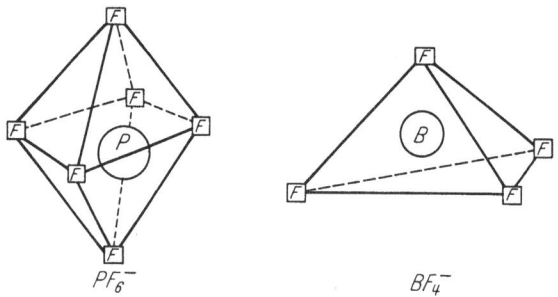

Fig. 5. Structure of the ions

cc) Polarizability. Meerwein (*33*) is of the opinion that the polarizability of the anion is the most important factor in determining the stability of an oxonium salt. The more difficult it is to polarize the halo ligand, the greater is the stability of the gegenion. In general, polarizability increases with increasing size and decreases with increasing charge on the central atom. Because P and B are smaller than Sb, PF_6^- and BF_4^- should be less polarizable than $SbCl_6^-$. But because of the greater charge on the P and Sb atoms, PF_6^- and $SbCl_6^-$ should be less polarizable than BF_4^-. Meerwein seems inclined to weigh charge somewhat more strongly and from his study of trialkyloxonium salts, in which he finds the $SbCl_6^-$ salts to be stable at higher temperatures than the BF_4^- salts, and arguments similar to those above, he concludes that $SbCl_6^-$ is more stable than BF_4^-. From their study of "non-tertiary" oxonium ions with differing gegenions, and a consideration of energies of formation, Klages, Meuresch, and Steppich (*59*) have also concluded that oxonium ions with an $SbCl_6^-$ gegenion are more stable than those with a BF_4^- gegenion.

Meerwein did not study PF_6^- but by analogous arguments the hexafluorophosphate ion might be expected to be the most stable of the three.

dd) Comparison of the stability of SbF_6^- and PF_6^-. Lange and Askitopoulos (*60*) present some data on the relative stabilities of hexafluoro complexes derived from Group V elements. They found that the corresponding hexafluorophosphates are generally more stable toward thermal and chemical influences than the hexafluoroantimonates. For example, the alkali salt of the antimony complex is almost immediately hydrolyzed in water while the corresponding phosphorus complex is completely stable. Upon introduction into water, the antimony complex reacts as follows:

$$[SbF_6]^- \underset{F^-}{\overset{OH^-}{\rightleftharpoons}} [SbF_x(OH)_{6-x}]^- \underset{F^-}{\overset{OH^-}{\rightleftharpoons}} [Sb(OH)_6]^- \qquad (26)$$

In general, the antimony salts are more soluble than the phosphorus complexes. The explanations that LANGE and ASKITOPOULOS give for this observed difference are twofold. First, due to its larger radius, the central antimony ion is less shielded and reaction occurs more readily. Second, the larger radius leads to greater polarizability and hence lower stability. One could expect that when the bulky Cl replaces the smaller F, the polarizability would increase still more, and hence the stability would be $PF_6^- > SbF_6^- > SbCl_6^-$. Apparently there have been no studies using the SbF_6^- as a gegenion in THF polymerizations.*

ee) Thermal stability. One might expect some correlation between the thermal and chemical stabilities of the gegenions. If this is true, Table 6, which shows the decomposition temperatures of various aryl diazonium salts, shows that for the most part the PF_6^- diazonium salts have a stability that is greater than or equal to that of the corresponding BF_4^- salts.

Table 6. *Decomposition temperatures of aryl diazonium salts, °C*

Compound	$X = BF_4^-$	Ref.	$X = PF_6^-$	Ref.
4-ClPhN$_2^+$X$^-$	126	61	155	25
$\overline{X}$N$_2^+$⟨ ⟩—⟨ ⟩—N$_2^+$X$^-$	135-7	62	150	63
PhN$_2^+$X$^-$	121-2	64	118	63
CH$_3$PhN$_2^+$X$^-$	110*	64	112**	63

* Para isomer.   ** Ortho isomer.

It should be noted also that these complex salts have rather amazing stability in water, especially considering the great reactivity of the halides themselves. $PhN_2^+BF_4^-$ can be recrystallized from water if one works quickly. We have regularly recrystallized $4\text{-Cl-}PhN_2^+PF_6^-$ from water at 30° C. followed by cooling in ice without any special haste. It could be this resistance to aqueous attack that gives these complex ions an advantage over other types of catalysts in producing high molecular weight polymers.

Corresponding thermal data with trialkyloxonium salts would be useful.

* *Note added in proof:* A recent patent (Neth Appl. 6,509,888, Feb. 1966; Chem. Abstr. *65*, 828D [1966]) diseloses the use of a variety of SbF_5^- salts for THF polymerization.

ff) Conclusion. We conclude that although there are no definite experiments in the literature, there is strong evidence to support the order of stability we propose on the basis of polymerization results, and that the remarkable success of the PF_6^- gegenion in producing high molecular weight PTHF lies in its greater stability. In the case of the other gegenions, decomposition leads to $SbCl_5$ and BF_3 which would tend to lower the molecular weight by a combination of transfer and termination reactions. With other gegenions which are not included in the above discussion, such as $FeCl_4^-$ and $AlCl_4^-$, similar arguments lead to the conclusion that the stability is probably even less and it is not surprising that equilibrium conversions and high molecular weight polymers are not obtained in their presence. With the present inexact knowledge of the nature of the gegenion in the aluminium alkyl catalyst systems it is difficult to make any predictions about their relative stabilities.

e) Other reactions of the gegenion. Undoubtedly, there are other reactions leading to chain termination which the oxonium ion-gegenion complex can undergo. Not all the observations made so far can be explained in terms of the chemistry considered in the rest of Section III C. For example, it is difficult to explain the formation of free selenium in the SeF_6 catalyzed polymerization of THF (*47*). The rather complete termination that seems to occur in polymerizations using $SbCl_6^-$ is also not understood fully. There probably are also other reactions that have not yet been defined or noted. Much work still needs to be done on the chemistry of termination in THF polymerizations.

D. Transfer

Chain transfer reactions in THF polymerizations have not been considered until rather recently. Compounds known to be effective chain transfer agents include dialkyl ethers, orthoesters, and water. In addition, chain transfer to polymer and with gegenion is possible.

1. With acyclic ethers. In a study of THF polymerization using PF_6^- gegenions Dreyfuss' (*25*) show that in the presence of dialkyl ethers chain transfer occurs and continues to occur after equilibrium is reached. The ultimate conversion to polymer is not affected but the intrinsic viscosity of the polymer decreases with time (Fig. 6). The reaction involved is essentially the reverse of the initiation reaction with trialkyl oxonium salts (equation 5). In the case of transfer the dialkyl ether reacts with the propagating oxonium ion to give a trialkyl oxonium ion which has one long chain alkyl and two short alkyls derived from the ether.

$$\sim\!CH_2CH_2O^+ \rceil \;+\; R_2O \;\longrightarrow\; \sim\!CH_2CH_2\!-\!O(CH_2)_4\!-\!O\!\!\underset{R}{\overset{R}{\diagup}}^+ \tag{27}$$

Attack by THF on one of the alpha carbons of the short alkyl groups completes the chain transfer process:

$$\text{\textasciitilde CH}_2\text{CH}_2\text{-O}^{+}\begin{smallmatrix}R\\\\R\end{smallmatrix} + \underset{O}{\bigcirc} \rightarrow \text{\textasciitilde CH}_2\text{CH}_2\text{-OR} + R\text{-O}^{+}\bigcirc \qquad (28)$$

If a cyclic ether such as dioxolane or dioxane is substituted for the acyclic ether the result is copolymerization of the cyclic ether (25) (Fig. 6).

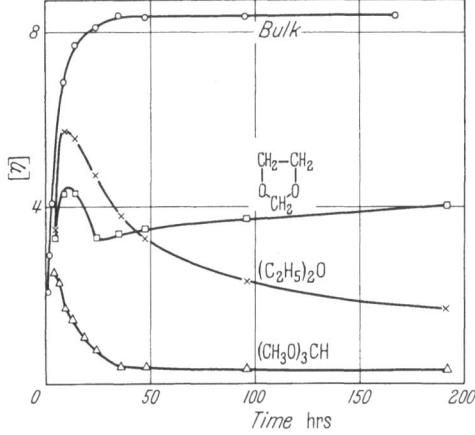

Fig. 6. Comparison of effect of additive on the intrinsic viscosity of PTHF as a function of time at 25° C. Additive (% by volume) and catalyst concentration (moles/l THF): (○) bulk, 5.1 × 10⁻³; (□) dioxolane —5.6%, 5.5 × 10⁻³; (×) ether —8.7%, 5.5 × 10⁻³; (△) TMOF —1.0%, 5.3 × 10⁻³ (25)

SIMS (40) obtained further evidence for transfer with diethylether by using C¹⁴ labeled ether as solvent. He showed that the activity of the polymer increased with time (Table 7).

Table 7. *Incorporation of radioactive ether in PTHF* ⋆ *[40]*

Polymerization time	$[\eta]$⋆⋆	$\dfrac{\text{moles Et}_2\text{O}}{\text{moles polymer}}$
3 hrs.	0.784	0.12
1 day	0.456	0.30
2 days	0.383	0.57

⋆ Initial monomer concentration = 6.3 moles/l [PF$_5$] = 4.5 × 10⁻² moles/l.
⋆⋆ Benzene at 20° C.

2. With orthoesters. Trimethyl orthoformate (TMOF) was found to be a very efficient chain transfer agent (25). Even one percent very markedly reduced the molecular weight of the PTHF (Fig. 6). In fact, it has been shown that the molecular weight of PTHF can be completely controlled by using TMOF. When [TMOF] > [cat], the molecular weight of the resultant polymer is dependent only on the concentration of TMOF (52).

Some insight into the unusually high efficiency of TMOF was obtained by allowing the polymerization of THF in the presence of a large amount of TMOF (40%) to proceed for 14 days. Under these conditions degradative chain transfer proceeded to the ultimate limit and only oligomers were isolated (Table 8). In this way it was clearly demonstrated that the end groups were both methoxy and that methyl formate was formed as a byproduct.

Table 8. *Products of polymerization of THF in 40% trimethylorthoformate (25)*

Compound isolated	Amt. from 41.1 g charge
$CH_3O(CH_2CH_2CH_2CH_2O)_nCH_3$	
$n = 1$	9.3
$n = 2$	4.6
$n > 2$ (Ave. 3.5)	7.6
$\underset{\substack{\parallel \\ }}{\text{O}}$ HCOCH$_3$	6.7
THF	10.0

Dreyfuss and Dreyfuss (25) explain these observations in terms of a transfer process which proceeds as follows:

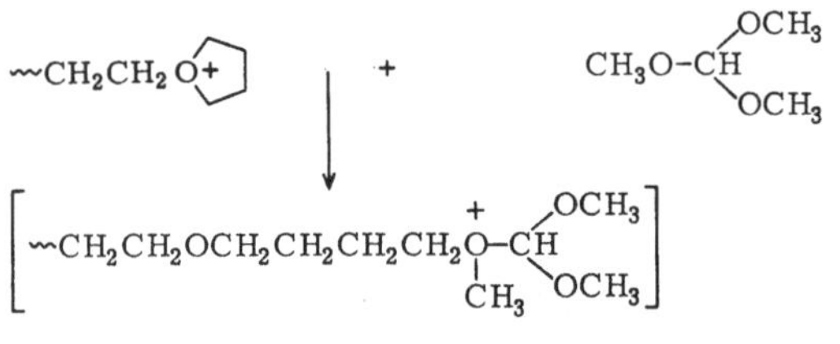

$$\sim CH_2CH_2 \overset{+}{\underset{\text{O}}{\bigcirc}} \quad + \quad CH_3O-CH\overset{OCH_3}{\underset{OCH_3}{\diagdown}} \tag{29}$$

$$\left[\sim CH_2CH_2OCH_2CH_2CH_2CH_2O-\overset{+}{\underset{\underset{CH_3}{|}}{C}}H\overset{OCH_3}{\underset{OCH_3}{\diagdown}}\right]$$

to form transiently an oxonium ion which leads to a methoxy end group and a resonance stabilized oxonium ion:

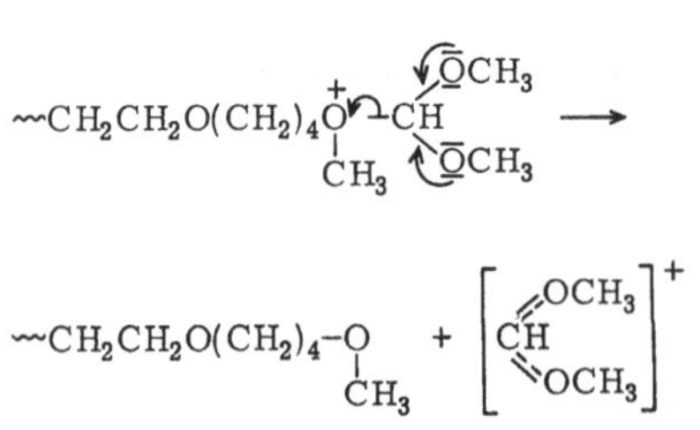

$$\sim CH_2CH_2O(CH_2)_4\overset{+}{O}\cdots CH\overset{OCH_3}{\underset{OCH_3}{\diagdown}} \longrightarrow \tag{30}$$

$$\sim CH_2CH_2O(CH_2)_4-\underset{\underset{CH_3}{|}}{O} \quad + \quad \left[CH\overset{OCH_3}{\underset{OCH_3}{\diagdown}}\right]^+$$

Although the above reactions are most easily visualized as presented, the reaction almost certainly proceeds in a concerted fashion as shown below.

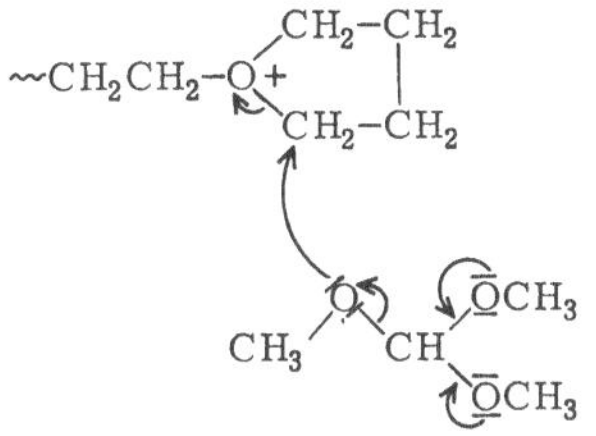

$$
\tag{31}
$$

When the chain transfer process is then completed in the following way (illustrated for clarity with a single resonance form),

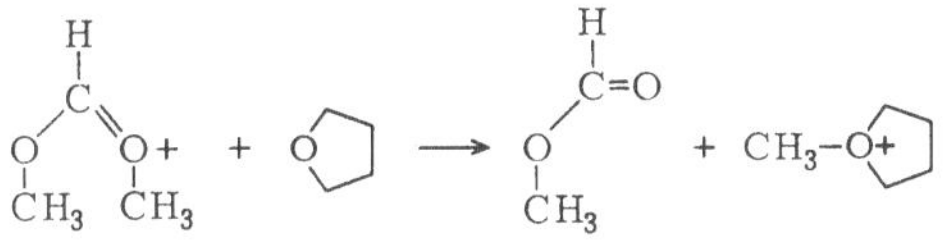

$$
\tag{32}
$$

a new chain having a methoxy group at the beginning to the chain is produced in addition to the observed methyl formate.

They explain the high degree of effectiveness of TMOF as a transfer agent by its ability to form a resonance stabilized oxonium ion.

3. With water. ROZENBERG et al. (37) have shown that in the $BF_3/$ ECH catalyzed polymerization of THF water can act as a chain transfer agent. They found that lower molecular weight polymers were produced when the amount of water added was less than the catalyst concentration. At the same time the polymerization rate remained the same. The mechanism they proposed for the reaction is as follows:

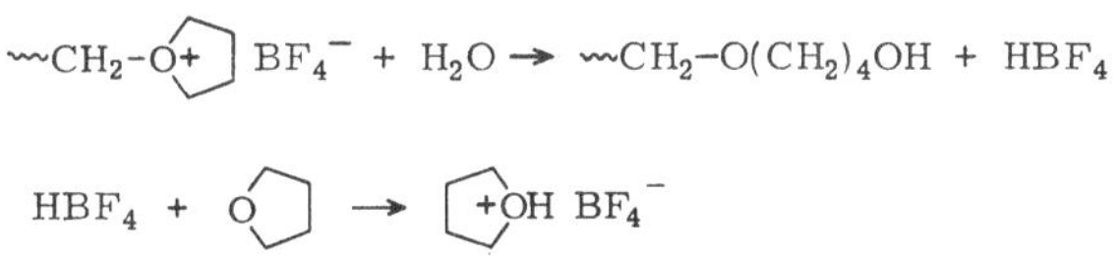

$$
\sim CH_2\text{-}O\!\!+\!\!\big] \; BF_4^- + H_2O \rightarrow \; \sim CH_2\text{-}O(CH_2)_4OH + HBF_4
$$

$$
HBF_4 + \; O\big] \rightarrow \big[\!+\!OH \; BF_4^-
$$

$$
\tag{33}
$$

When these reactions are written as in equation 34, the process becomes analogous to the reaction with dialkyl ether (equations 27 and 28).

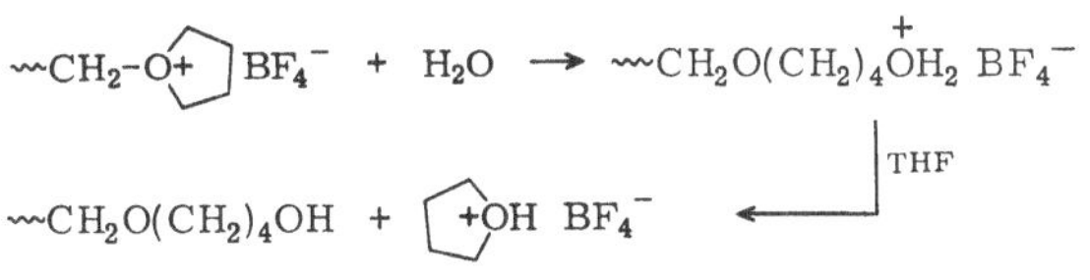

$$
\sim CH_2\text{-}O\!\!+\!\!\big] BF_4^- + H_2O \rightarrow \; \sim CH_2O(CH_2)_4\overset{+}{O}H_2 \; BF_4^-
$$

$$
\sim CH_2O(CH_2)_4OH + \big[\!+\!OH \; BF_4^- \longleftarrow \quad \Big| THF
$$

$$
\tag{34}
$$

SIMS (40) has studied the effects of water on PF_5 initiation of THF polymerization (Section IIIB2b). Possibly, in addition to cocatalysis and destruction of catalyst, his results are complicated by transfer.

Small amounts of alcohol should behave like water. Alcohol would enter in equation 34 to lead to dead chains with –OR end groups instead of –OH end groups. The ethylene glycol used by Murbach and Adicoff (67) in their copolymerization of THF with ethylene oxide appears to us to function as a transfer agent. They call ethylene glycol a cocatalyst but present data which show that a lower molecular weight polymer is produced when more ethylene glycol is charged. In this case the end group of the dead chains should be –OCH$_2$CH$_2$OH.

4. With polymer oxygen. The oxygen atoms in the polymer chain itself are really a kind of dialkyl ether and especially in the later stages of the polymerization can react in the same way (equation 27) to form an oxonium ion:

$$\text{~CH}_2\text{CH}_2\text{O+} \quad + \quad \text{O}\begin{smallmatrix} \text{CH}_2\text{CH}_2\text{~} \\ \\ \text{CH}_2\text{CH}_2\text{~} \end{smallmatrix} \quad \rightarrow \quad \text{~CH}_2\text{CH}_2\text{O(CH}_2)_4\text{O+}\begin{smallmatrix} \text{CH}_2\text{CH}_2\text{~} \\ \\ \text{CH}_2\text{CH}_2\text{~} \end{smallmatrix} \qquad (35)$$

However, the product is still a reactive species and can regenerate the propagating species by reaction with monomer. The consequences of this reaction have been treated mathematically by Hermans (50). He discussed the chain length distribution in a polymer such as PTHF in which chain ends react at random with all monomer units and how this distribution changes with time. He points out that it is intuitively obvious that this reaction will eventually lead to a geometric distribution of chain lengths (the "most probable distribution"). Both the total number of monomer units and the total number of molecules remain constant. Hermans finds that the approach to the steady state value will occur more rapidly with polymers of high molecular weight than with low molecular weight polymers. Thus it is seen that this reaction is a type of transfer reaction that leads only to a redistribution of molecular weight.

Rozenberg et al. (51) used the reaction shown in equation 35 to explain the large increase in the viscosity of the reaction mixture that they observed after most of the monomer had polymerized. As evidence they offered the observation that the addition of an amount of methanolic KOH comparable to the number of active centers caused a marked reduction of viscosity. They associated this with a breakdown of the product of equation 35. Sims (37) has made similar observations. He suggests that the breakdown may be associated with a hydrolysis of the phosphate linkages in the chain of the final product shown in equation 8. Equation 6a suggests that the explanation for Rozenberg's observations may be similar to Sims, i. e. a hydrolysis of borate esters occurs.

Attack by a polymer oxygen could also lead to the formation of a macrocyclic oxonium ion (equation 36):

$CH_2CH_2O^+$
CH_2
CH_2
O
CH_2
CH_2
$CH_2CH_2OCH_2CH_2$
CH_2
CH_2
O
+
CH_2
*
CH_2
CH_2
CH_2

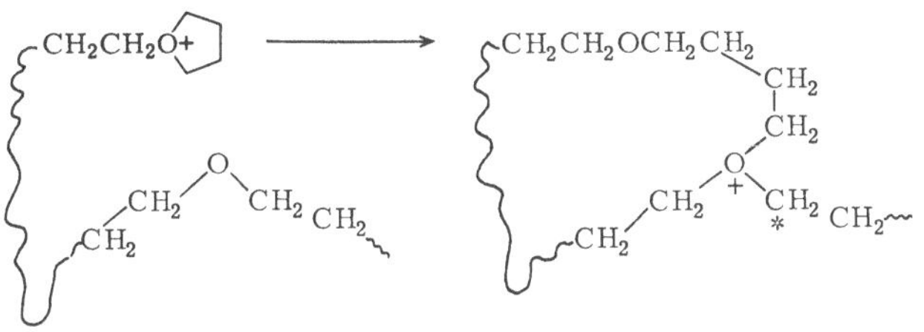

(36)

In this case, however, attack by monomer at the carbon indicated with an asterisk would lead to a macrocyclic oligomer and a shortened propagating oxonium ion. This would be a true transfer reaction in that a smaller "dead" molecule is produced without any loss of active centers.

Rose (*46*) has reported the formation of a cyclic tetramer in the polymerization of oxetane and in fact under some conditions this 16 membered ring is the major product. The tetramer may be a depropagation product and the lowest energy ring which can be formed. In the cationic polymerization of ethylene oxide in addition to polymer, a large amount of dioxane is formed (*65*). Eastham (*65*) suggests that some of the dioxane forms by a separate mechanism but that much of it results from the depropagation reaction which leads to a stable six-membered ring. In the polymerization of THF, depropagation leads only to monomer and the formation of larger cyclic species has not been reported. The five-membered ring of the monomer is probably the most stable ring possible. Thus, we see that reaction of the type indicated in equation 36 is probably unimportant.

5. With gegenion. A form of chain transfer may also occur by reaction with gegenion. For a general Lewis acid gegenion, MtX_{n+1}^-, the reaction would be:

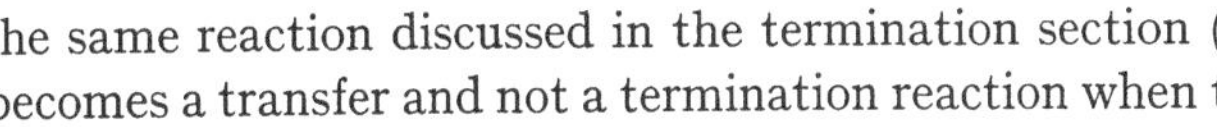

CH_2O^+
+ MtX_{n+1}^- →
$CH_2O(CH_2)_3CH_2X$ + MtX_n

(37)

This is the same reaction discussed in the termination section (equation 22) but becomes a transfer and not a termination reaction when the MtX_n formed is capable of generating new chains by itself. Such a chain transfer process has not yet been clearly demonstrated but might be possible in the case of $SbCl_6^-$ or PF_6^- gegenions since $SbCl_5$ and PF_5 initiate polymerization without added promotor.

6. Other. In the absence of solvents and using PF_5 catalyst, Sims (*26*) has reported a fall of intrinsic viscosity with time (Fig. 7) similar to that observed by Dreyfuss' (Fig. 6) in the presence of added transfer agents.

This type of fall of molecular weight with time in a system which probably does not suffer from significant termination reactions suggests chain transfer with some agent inadvertantly present.

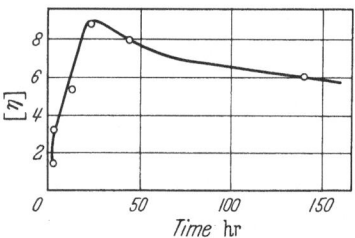

Fig. 7. Change of intrinsic viscosity with time of PTHF polymerized in bulk at 20° C using PF$_5$ catalyst; ([PF$_5$] = 0.762 g/l) (26)

Bawn, Bell, Fitzsimmons, and Ledwith (20) have suggested that transfer reaction in the bulk polymerization of THF must involve either hydride ion abstraction from the alpha methylene of THF or of tetra-methyleneoxy units in the polymer, or degradative oxonium ion formation with the ethereal oxygen atoms of polymers of the type discussed in Section III D 4.

IV. Kinetics

Several kinetic studies of the polymerization of THF have been reported. Nearly all of the workers conclude from their studies that the system is "living". Sometimes the evidence given is that on addition of more monomer after equilibrium is reached, polymerization continues at the same rate within experimental error. It is frequently noted that the rate of polymerization increases linearly as the catalyst concentration increases. In most cases no direct measurement of the number of active centers has been possible. As we have shown above, from the failure of many of the systems to reach equilibrium conversion or from the fall in viscosity that occurs after equilibrium conversion is reached, most of these systems suffer from some termination and/or transfer reactions. However, in the early stages of the polymerization and at low temperatures (around 0° C) THF appears to give a kinetically "living" polymerization with some catalysts. The number of active centers remains constant either from the beginning of the reaction or after a certain induction period. Since many differences arise from the catalyst system used, we have grouped the studies according to catalyst.

A. Triethyloxonium tetrafluoroborate

Of all the kinetic studies that have appeared Vofsi and Tobolsky (32) have carried out the most quantitative study. In their work they used preformed triethyloxonium tetrafluoroborate, $(C_2H_5)_3O^+BF_4^-$, as catalyst and carried out the polymerizations at $0°$ C in a solution of dichloroethane. By using C^{14} labeled catalyst, they were able to show that the catalyst concentration could be equated to the number of active centers and to measure the rates of initiator disappearance. VOFSI and TOBOLSKY found that the catalyst conversion was over 90% when monomer was polymerized to the extent of about 15%. Thus they were able to determine the degree of propagation precisely in the early stages of the polymerization.

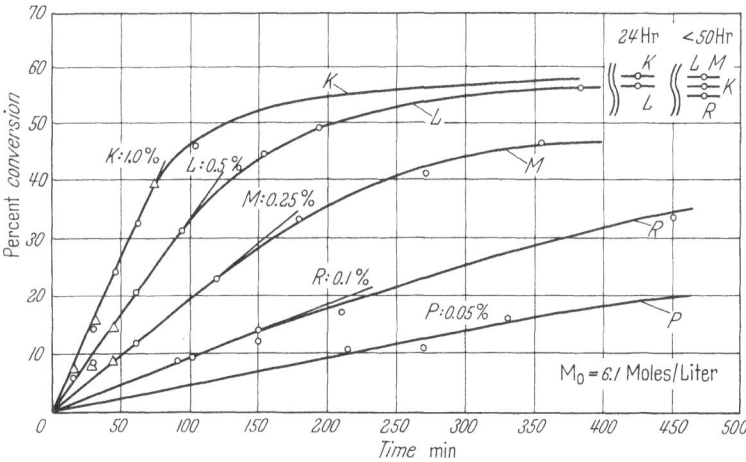

Fig. 8. Conversion vs. time plot at $0°$ C. The letters and numbers designate run and percentage of catalyst to monomer: run K, 6.1×10^{-2} mole/l (1% of monomer concentration); run L, 3.05×10^{-2} mole/l (0.5%); run M, 1.53×10^{-2} mole/l (0.25%); run R, 0.61×10^{-2} mole/l (0.1%); and run P, 0.31×10^{-2} mole/l (0.05%). Triangles indicate duplicate runs (32)

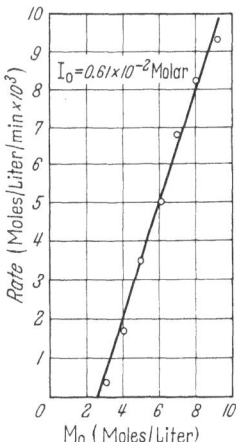

Fig. 9. Initial rate of polymerization versus initial monomer concentration (32)

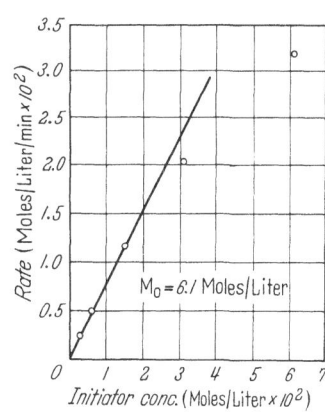

Fig. 10. Initial rate of polymerization vs. initial catalyst concentration (32)

They studied the rate as a function of both catalyst concentration and initial monomer concentration and obtained the results shown in Figs. 8, 9, and 10. Initial rates are directly proportional to initial monomer concentrations (Fig. 9). But as can be seen from Fig. 10 initial rates begin to deviate from linearity above a catalyst concentration of 2×10^{-2} mole/l. The rate of initiator disappearance was given by:

$$-d[I]/dt = k_i[I]([M] - [M_e]) . \qquad (38)$$

They found that the rate constant, k_i in equation 38, was 1.3 to 1.7×10^{-2} for initial catalyst concentrations in the range 6.1×10^{-2} to 1.53×10^{-2} mole/l, $[M_e]$ of 2.62 mole/l, and $[M_0]$ of 6.1 mole/l. The data as a whole seemed to support the mechanism proposed above (Section III B 1) for initiation with oxonium salts, namely:

$$
\begin{array}{c}
\text{Et} \\
\diagdown \overset{+}{\text{O}}\text{-Et BF}_4^- \ + \ \text{O}\!\!\bigcirc \ \xrightleftharpoons{k_i} \ \overset{\text{Et}}{\diagdown}\!\!\text{O} \ + \ \text{Et-}\overset{+}{\text{O}}\!\!\bigcirc \text{BF}_4^- \\
\text{Et} \hspace{5.5cm} \text{Et}
\end{array} \qquad (39)
$$

$$
\text{Et-}\overset{+}{\text{O}}\!\!\bigcirc \ \text{BF}_4^- \ + \ \text{O}\!\!\bigcirc \ \xrightleftharpoons[k_d]{k_p} \ \text{Et-OCH}_2\text{CH}_2\text{CH}_2\text{CH}_2\text{-}\overset{+}{\text{O}}\!\!\bigcirc \ \text{BF}_4^- \qquad (40)
$$

where k_i, k_p, k_d are the specific rate constants of initiation, propagation, and depropagation, respectively. Furthermore, their data could be interpreted in terms of the rate of reaction for an equilibrium polymerization without termination (32):

$$-dM/dt = k_i[I]([M] - [M_e]) + k_p([I_0] - [I])([M] - [M_e]) . \qquad (41)$$

Within a monomer concentration range of $4-10$ mole/l and a catalyst concentration below 3.0×10^{-2} moles/l, initiation is fast, and the initial rate is directly proportional to $[I_0]$. Thus $[I]$ is very small, and the rate is closely approximated by:

$$-dM/dt = k_p[I_0]([M] - [M_e]) \qquad (42)$$

or in its integrated form:

$$2.3 \log\{([M_0] - [M_e])/([M] - [M_e])\}/[I_0] = k_p t . \qquad (43)$$

A plot of equation 43 is shown in Fig. 11. From it a value of k_p of 0.290 l/min mole (4.83×10^{-3} l/mole sec) was deduced. Vofsi and Tobolsky conclude, however, that there is little doubt that to cover a wider kinetic range, two constants, namely k_i and k_p, are insufficient, and a termination term would have to be included to more fully describe the system.

Rozenberg et al. (24) have also studied the kinetics of polymerization of THF initiated by triethyl oxonium tetrafluoroborate. They generated their catalyst in situ from epichlorohydrin and the boron trifluoride-ether complex and carried out their polymerizations in bulk and in diethyl ether solution. Both DREYFUSS and DREYFUSS (25) and SIMS (40) have shown that transfer reactions occur in diethyl ether. ROZENBERG et al. apparently did not observe any transfer in the relatively short time of their studies (up to five hours). They state that their experimental data fit equation 42. They define $[I_0]$ in terms of the catalyst components charged assuming instantaneous and quantitative conversion to catalyst.

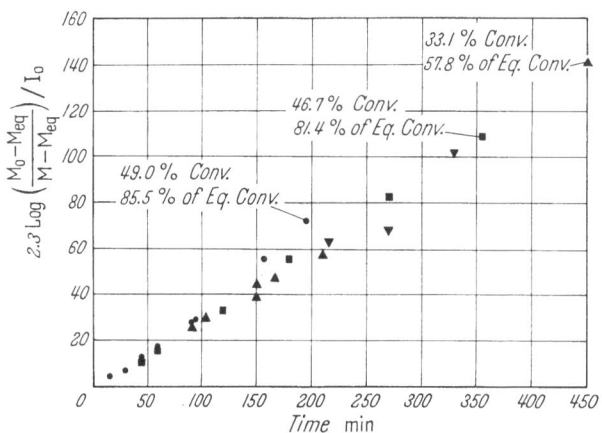

Fig. 11. Integrated form of conversion vs. time curves for runs L, M, R, and P (compare Fig. 8) (32)

The rate constant of polymerization that they deduced for THF at 20° C was 1.66×10^{-2} l/mole sec. From studies of the polymerization in the temperature range $0-40°$ C they derived an activation energy, E_a, of 13.3 kcal/mole and a pre-exponential factor, A, of 1.64×10^8 l/mole sec. Using these values for E_a and A we have calculated a value of 3.7×10^{-3} l/mole sec for k_p at 0° C. This is in good agreement with the value, 4.83×10^{-3} l/mole sec, reported by VOFSI and TOBOLSKY (32).

By using the value of k_p at 20° C and the value of the equilibrium monomer concentration, the rate constant calculated for the depropagation reaction, k_d, was 4.67×10^{-2} sec^{-1} at 20° C. An activation energy of 19.4 kcal/mole and a pre-exponential factor of 1.65×10^{13} sec^{-1} were calculated for the depropagation reaction.

ROZENBERG et al. (31) found in their studies that the molecular weight of the polymer they obtained was higher than would be expected from the ratio of the monomer charged to the catalyst charged. VOFSI and TOBOLSKY (32) did not notice this discrepancy in their work with the preformed oxonium salt.

B. Triethyloxonium hexachloroantimonate

LYUDVIG, ROZENBERG et al. (41) made a study of the kinetics of THF polymerization initiated by triethyloxonium hexachloroantimonate.

They carried out their measurements at 20° C in bulk. They report (31) that in this case it was necessary to use the preformed salt. When the catalyst was added in the form of two components, initiation was not complete. The kinetics were analogous to those discussed above. An activation energy of 15.0 kcal/mole was derived. Equation 42 was followed. The plot of $\log(M_0 - M_e)/(M - M_e)$ against time (compare equation 43) is shown in Fig. 12.

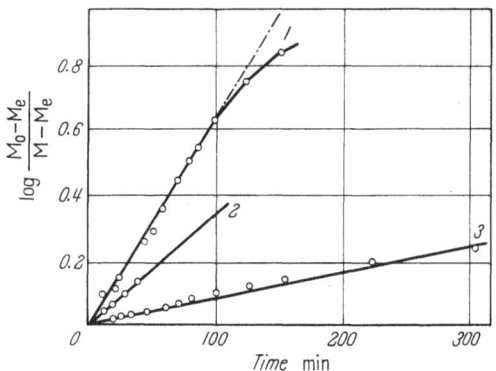

Fig. 12. Time dependence of $\log[(M_0 - M_e)/(M - M_e)]$ in polymerization of THF in the presence of $(C_2H_5)_3$ $O^+SbCl_6^-$ at 20°. $[M_0] = 12.3$ mole/l; 20°; $[C_0]$ (mole/l): 1 — 1.3 × 10⁻³; 2 — 0.59 × 10⁻³; 3—0.18 × 10⁻³ (41)

C. Triethylaluminum-water-promotor system

Imai, Saegusa, Furukawa et al. (48, 66) carried out kinetic studies of THF polymerization in bulk and in cyclohexane solution at 0° C. They used a ternary catalyst system consisting of $AlEt_3$–$H_2O(2:1)$-epichlorohydrin (ECH). They obtained high molecular weight polymer and noticed no evidence for either termination or transfer. Their polymerizations were preceded by an induction period as shown in Fig. 13 but after that their data could be fitted to an equation of the same form as equation 42. This time $[I_0]$ was defined as the concentration of propagating species ($[P^*]$) determined from the amount and the molecular weight of product polymer.

The concentration of propagating species was found to be proportional to the amounts of $AlEt_3$ and of ECH used, so equation 42 now takes the form:

$$-dM/dt = k_p\{k'[ECH]_0[AlEt_3]_0\}\{[M] - [M_e]\} \tag{44}$$

where initiator concentration is represented by:

$$[I_0] = [P^*] = k'[ECH]_0[AlEt_3]_0. \tag{45}$$

These workers found this relationship to hold as long as $[ECH]_0/[AlEt_3]_0 < < 0.07$. Beyond this region the linear relationship did not hold and the values of $[I_0]$ were lower than could be expected from the linear relationship. When $[ECH]_0/[AlEt_3]_0 > 0.3$, $[P^*]$ was no longer affected by the initial concentration of ECH. They also observed, as shown in Fig. 14, that at early conversions the molecular weight was higher than would be expected from the catalyst charged. The rate constants, k_p, derived by these workers are 1.28×10^{-2} l/mole sec and 1.40×10^{-2} l/mole sec for the bulk and solution polymerizations respectively.

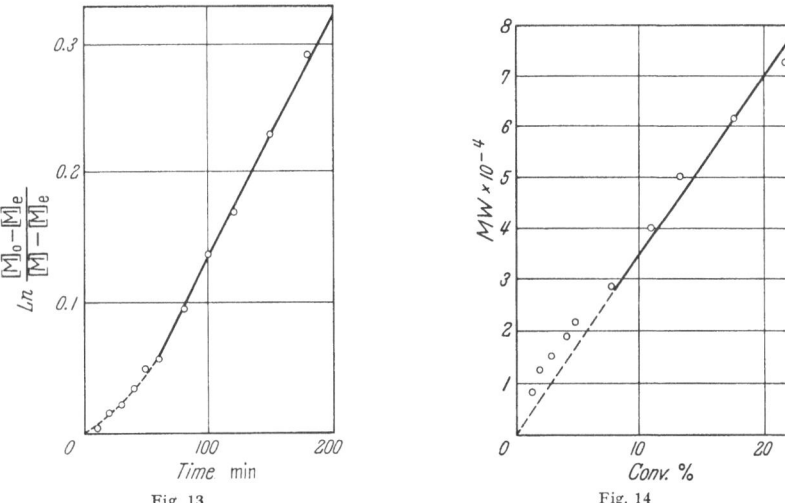

Fig. 13   Fig. 14

Fig. 13. First order plot of monomer concentration, bulk polymerization at $0°$ C, $[M]_0 = 12.1$ mole/l, $[M]_e = 2.63$ mole/l, $[AlEt_3] = 0.179$ mole/l, $[H_2O]/(AlEt_3) = 1/2$ $[ECH]_0 = 0.159$ mole/l (48)

Fig. 14. Molecular weight versus conversion, bulk polymerization at $0°$ C, $M_0 = 12.1$ mole/l, $[AlEt_3] = 0.179$ mole/l, $[H_2O]/[AlEt_3] = 1/2$, $[ECH]_0 = 0.159$ mole/l (48)

D. Phosphorus pentafluoride-promotor and phosphorus pentafluoride systems

Sims $(37, 40)$ has studied the polymerization of THF in ether solution initiated either by PF_5–ECH or PF_5.

The results of his studies with the PF_5–ECH system (37) are similar to those of Imai, Saegusa, Furukawa et al. $(48, 66)$. Once again the rate of polymerization was first order with respect to [ECH] up to a limit, in this case $1/4$ $[PF_5]$ (Fig. 15), and first order with respect to $[PF_5]$ at constant $[PF_5]/[ECH]$ concentration. Sims data again corresponded to equation 42, where in this case $[I_0]$ was the active species formed by reaction of PF_5 and ECH in the proportion of about $4:1$. More specifically he reports that

$$-dM/dt = 0.85 \, [ECH] \, \{[M] - 1.6\} \, \text{mole/l min}, \quad [PF_5] > 4 \, [ECH] \quad (46)$$

and

$$-dM/dt = 0.21 \, [PF_5] \, \{[M] - 1.6\} \, \text{mole/l min}, \quad [ECH] \geqq 1/4 \, [PF_5]. \quad (46a)$$

From polymerizations at 8, 14, and 20° C. Sims derives an activation energy of 14.5 kcal/mole.

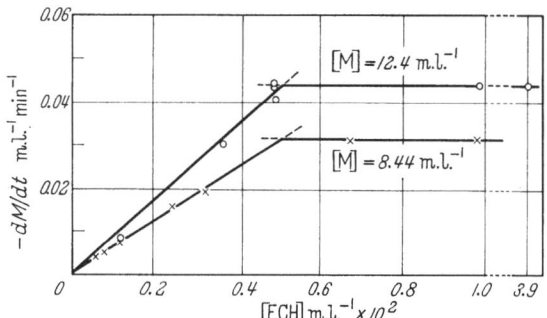

Fig. 15. Effect of ECH on polymerization of THF by PF_5 at 20° C. $[PF_5] = 1.9 \times 10^{-2}$ mole/l (37)

PF_5 gas as a catalyst in the absence of ECH has given more complex kinetics. Sims (40) found that the rate of monomer disappearance at any given time was higher for higher initial monomer concentrations. There

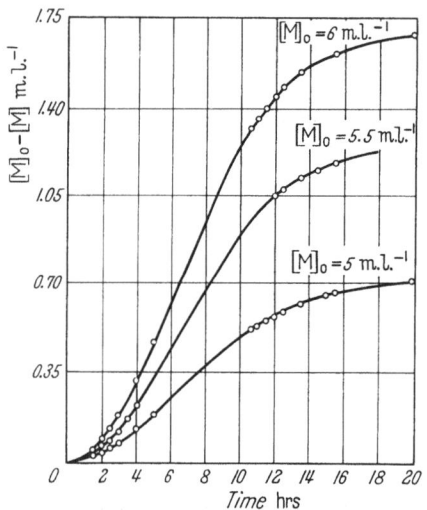

Fig. 16. Polymerization of THF by PF_5 at 20° C. $[PF_5] = 1.9 \times 10^{-2}$ mole/l (40)

was apparently an induction period characteristic of slow initiation (Fig. 16). Sims' results closely fit the relationship

$$(H - H_0)^{1/2} = Kt \qquad (47)$$

where $(H - H_0)$ represents the observed change in dilatometric height, and t is time. He also found a marked dependence on the concentration of

added water. The rate increased to a maximum at an added water to
PF$_5$ ratio of about one to six. The rate then decreased until polymerization no longer occurred when the water to PF$_5$ ratio was 1.3 to one or
greater.

E. Antimony pentachloride

LYUDVIG, ROZENBERG et al. (41) have also considered the kinetics of
polymerization of THF initiated by antimony pentachloride. Their rate
curves are clearly S-shaped (Fig. 17). This is similar to the results obtained in the triethyl aluminum system (Fig. 13) and in the PF$_5$ case (Fig.16).

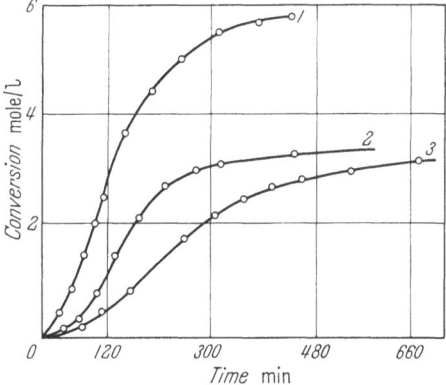

Fig. 17. Polymerization of THF by SbCl$_5$ at 20° C. 1 — [M_0] = 8 mole/l, [C_0] = 0.038 mole/l, 2 — [M_c]
= 6 mole/l, [C_0] = 0.0668 mole/l, 3 — [M_0] = 6 mole/l, [C_0] = 0.0334 mole/l (41)

LYUDVIG, ROZENBERG et al. (41) interpreted this to mean that the rate of
initiation is slow. In fact, they found that in this case the rate of initiation
is so slow that even at the end of the reaction, only a small fraction of

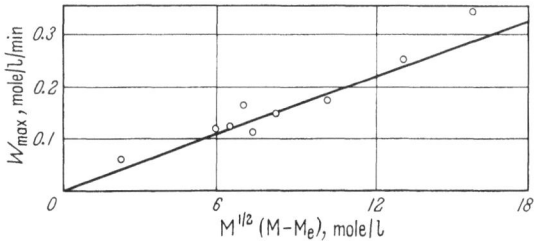

Fig. 18. Dependence of maximal rate on monomer concentration in polymerization of THF in the presence of
SbCl$_5$. [C_0] = 0.0334 mole/l (41)

initiator has been consumed and a steady-state process never develops.
The data appear to fit an equation of the form

$$ -dM/dt = (k_i\, k_2\, k_p)^{1/2}\, [C]\, [M]^{1/2}([M] - [M_e])\,. \tag{48} $$

The experimental results are shown in Fig. 18. It is this kinetic evidence which is given to support the mechanism of initiation proposed by these workers (compare equation 15).

F. Other initiators

Lyudvig, Rozenberg et al. (41) have reported a study of the kinetics of polymerization of THF in the presence of the reaction product of CH_3COCl and $SbCl_5$, which they write as $CH_3CO^+SbCl_6^-$. The rate curves were S-shaped but there was a point of maximum rate and a steady-state was reached. The activation energy, 15 kcal/mole, was the same as they found for THF polymerization with $Et_3O^+SbCl_6^-$ initiation.

Bawn, Bell, and Ledwith (17) have reported a study of the kinetics of the polymerization of THF catalyzed by $Ph_3C^+SbCl_6^-$. The activation energy they obtained is 12 kcal/mole. Bawn, Bell, Fitzsimmons, and Ledwith (20) have compared the rates of polymerization initiated by $Ph_3C^+SbCl_6^-$ with those initiated by $Ph_3C^+PF_6^-$. Their results are shown in Table 9. They conclude that there is not much difference in the rate of polymerization as a result of replacing $SbCl_6^-$ by PF_6^-.

Table 9. *Bulk polymerization of THF using triphenyl methyl carbonium ion salts as initiators* (20)

Gegenion X^-	$Ph_3C^+X^-$ 10^3 mole/l	Temp. °C	$10^4 R_p$ mole/l sec
PF_6^-	2.14	25	2.64
$SbCl_6^-$	2.14	25	1.11
PF_6^-	3.82	25	4.47
$SbCl_6^-$	3.82	25	1.91
PF_6^-	4.52	50	15.8
$SbCl_6^-$	4.52	50	10.6
PF_6^-	6.65	50	24.8
$SbCl_6^-$	6.65	50	15.4

G. Conclusion

From these studies it appears that the kinetics of polymerization of THF are closely approximated by equation 42. The equation does not always apply from the beginning of the polymerization and frequently cannot be applied before a steady-state of active centers is achieved. The initiator term, I_0, in this equation is often a function of several components. Only in the case of preformed trialkyloxo nium ions of the form $R_3O^+X^-$ is the initiation simple. These results suggest that in order to theoretically study the kinetics of polymerization of THF or to compare the kinetics of THF polymerization in the presence of different gegenions, it is desirable to use preformed trialkyl oxonium salts. Ideally

an alkyl tetrahydrofuranyl salt, ⟨⟩+O-RX⁻, should be used but practically the triethyl oxonium salts are probably preferable. To date no
effort has been made to measure either a transfer rate, a termination
rate, or a depropagation rate. When such studies have been made, we
should have a greater understanding of all the factors affecting the
polymerization of THF and other cyclic ethers. Of course, kinetic studies
will continue to be useful with the various Lewis acid-promotor catalysts
as an aid in deducing the mechanism of initiation.

V. Properties

Polytetrahydrofuran is a crystalline polymer. It can be prepared at
any desired molecular weight ranging from oligomers containing only
a few monomers up to polymers with a degree of polymerization in the
tens of thousands. At the lower end of the molecular weight scale the
products are sticky viscous oils at room temperature. In the intermediate
regions they first become waxes that flow easily on melting, and then
with further increase in molecular weight they become tough, rubbery
materials that retain their shape on melting. At very high molecular
weights they are almost intractable. Quite a variety of studies of the
properties of PTHF, many of them preliminary in nature, have appeared
in the last two to three years.

A. Solution properties

1. Intrinsic viscosity. Viscosity is the property most commonly cited
in the characterization of PTHF. The majority of authors have quoted
values determined in benzene, usually at 25° C. However, as shown in
Table 10, there is a wide variety of other solvents in use. Most of the
studies have been made on low molecular weight polymers. When higher
molecular weights were used, some efforts to fractionate the polymers have
been made. For example, MAKLETSOVA, EPEL'BAUM, ROZENBERG, and
LYUDVIG (70) separated PTHF into fractions by successive precipitation
from toluene with methanol. KURATA, UTIYAMA, and KAMADA (69)
assumed their distributions were narrow but checked one sample by
analysis of the sedimentation velocity in a theta solvent and found an
$\overline{M_w}/\overline{M_n}$ ratio of 1.04 for this sample. ALI and HUGLIN (68) collected
fractions after incremental addition of MeOH. The variation in the
resulting Mark-Houwink equations that have been published likewise is
illustrated in Table 10.

Table 10 shows that there is disagreement on the molecular weight
to be associated with a given intrinsic viscosity. For example, an intrinsic
viscosity of **3.8** measured in benzene at $25-30°$ C gives values of molecular
weight ranging from 150,000 to 600,000. Thus it seems that it is not at

present advisable to calculate a molecular weight from a given intrinsic viscosity for the purpose of determining catalyst efficiency or the number of active growing centers. Further work is needed to establish the effect of molecular weight, and molecular weight distribution on intrinsic viscosity. Perhaps the polymers produced by different catalysts are not identical.

Table 10. *Mark-Houwink equations for PTHF*

| Solvent | Temp. °C | M. W. Range × 10⁻⁴ | [η] = KMα | | | Ref. |
			M	$K \times 10^4$	α	
toluene	28	3.2—11.8	$\overline{M}_n$	2.51	0.78	68
toluene	30	0.18—1.26	$\overline{M}_n$	2.36	0.777	34
benzene	25	—	$\overline{M}_n$	2.98	0.79	20
benzene	20	"low"	$\overline{M}_n$	2.5	0.82	40
benzene	30	3.5—110	$\overline{M}_w$	13.1	0.60	69
cyclohexane	30	3.5—110	$\overline{M}_w$	17.6	0.54	69
ethyl acetate	30	3.5—110	$\overline{M}_w$	4.22	0.65	69
theta mixture *	31.8	3.5—110	$\overline{M}_w$	34.3	0.45	69
theta solvents **	**	—	M	23.0***	0.5***	71
methyl ethyl ketone	25	1.7—150	$\overline{M}_w$	21.0	0.5	24, 70

   * Ethyl acetate (22.7 wt.%)/n-hexane (77.3 wt. %).
  ** Isopropyl alcohol at 44.6° C, diethyl malonate at 33.5° C, ethyl acetate (22.3 wt. %)/n-hexane (77.7 wt. %) at 30.5 ± 1° C, n-butanol at 5.0 ± 1° C.
 *** Average value for the four theta solvents.

2. Effect of solvent and temperature on intrinsic viscosity. Kurata, Utiyama, and Kamada (*69*) have compared the intrinsic viscosity of PTHF in several different solvents at the same temperature. Some of their results are shown in Table 11.

Table 11. *Effect of solvent on intrinsic viscosity (dl/g) of PTHF (69)*

Sample	Ethyl acetate 30° C	Benzene 30° C	Cyclohexane 30° C	Theta mixture 31.8° C
K-124	0.402	0.700	0.473	0.367
K-101	0.514	0.912	0.590	0.440
K-106	1.20	1.836	1.125	0.828
K-120	2.07	3.31	1.900	1.115
K-110	2.80	4.25	2.75	1.352
K-117	3.21	5.38	—	1.590

Viscosities in benzene, the most commonly used solvent, tend to be higher than in the other solvents and increase very rapidly with increasing molecular weight. This suggests that benzene is too good a solvent. A poorer solvent such as ethyl acetate would be a wiser choice.

3. Theta solvents. Selection of a poor solvent for a polymer is desirable when making solution property measurements because it permits the use of higher concentrations and minimizes the effects of nonideality. The most suitable choice is a theta solvent (73). Table 12 lists the theta solvents and the corresponding theta temperatures which have been found for PTHF.

Table 12. *Theta solvents for PTHF*

Theta solvent	Theta temp. °C	Ref.
ethyl acetate (22.7 wt. %) and n-hexane (77.3 wt. %)	31.8	69
ethyl acetate (22.3 wt. %) and n-hexane (77.7 wt. %)	30.5 ± 1	71
n-butanol	5.0 ± 1	71
diethyl malonate	33.5	71
heptane and cyclohexane	26	72
isopropyl alcohol	44.6	71

4. Molecular weight determination.

a) Number average molecular weight. Low number average molecular weights have usually been determined by end group analysis for hydroxyl

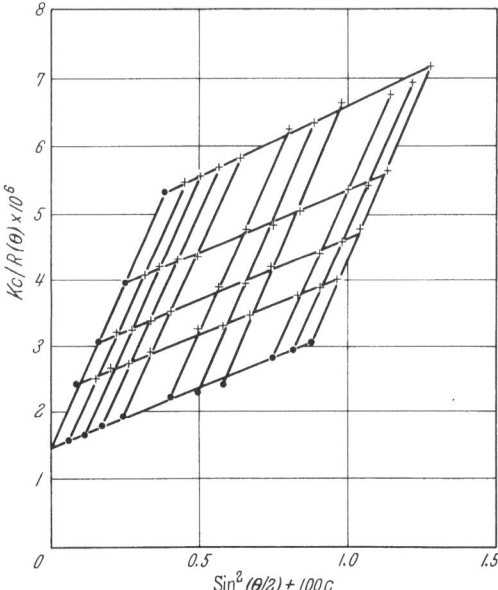

Fig. 19. Zimm plot for light scattering data on PTHF in ethyl acetate at 30° C (69), [η] in ethyl acetate = 2.80, M_W = 688,000

groups using titration (34, 75), infrared analysis (40), or ebulliometry (74). VOFSI and TOBOLSKY (32) determined number average molecular weights in a Mechrolab 301-A vapor pressure osmometer using benzene as solvent.

They also carried out end group analysis by means of a Chicago Nuclear liquid scintillation counter to count C^{14}-labeled end groups.

Higher number average molecular weights have been determined by osmometry, most frequently in benzene or toluene (25, 68, 74).

b) Weight average molecular weight. PTHF apparently gives no trouble in light scattering studies. MAKLETSOVA, EPEL'BAUM, ROZEN-BERG, and LYUDVIG (70) determined the weight average molecular weight of PTHF in methyl ethyl ketone and chlorobenzene. They used a visual nephelometer at three angles to the primary beam. KURATA, UTIYAMA and KAMADA (69) used ethyl acetate and a mixed theta solvent, ethyl acetate (22.7 wt.-%) and n-hexane (77.3 wt.-%). Their measurements were made using a Brice type photometer at various scattering angles between 30 and 140°. Apparently PTHF is capable of producing very nice Zimm plots without curvature. One of KURATA et al.'s plots is shown in Fig. 19.

Both sets of workers studied a wide range of molecular weights from about 30,000 to 1,000,000. The refractive index increments, dn/dc and the refractive indexes, n_D, for their solvents are given in Table 13.

Table 13. *Refractive index and refractive index increments for light scattering solvents*

Solvent	n_D	dn/dc	λ, mμ	$A_2 \times 10^4$ mole g^{-2} ml	Ref.
theta solvent *	1.3684	0.113$_9$	436	0, —0.0785, —0.0985**	69
ethyl acetate	1.3698	0.113$_6$	436	2.47 to 7.35	69
chlorobenzene	1.52479	0.070	436	—	70
methyl ethyl ketone	1.38071	0.095	546.1	—	70

* Ethyl acetate (22.7%)/n-hexane (77.3%).
** Values were determined for three different polymers.

c) Molecular weight distribution. Studies of the molecular weight distribution and its changes during a "living" polymerization of PTHF offer great promise of providing new experimental information that has not been readily obtained with other systems. BROWN and SZWARC (76) originally pointed out that a normal ("most probable") distribution of molecular weights will ultimately result from polymerization of a "living" polymer. However, if the depropagation reaction is slow, and initiation is rapid a narrow (POISSON) distribution will form initially. MIYAKE and STOCKMAYER (77) have recently investigated the mathematical problem of a reversible living polymer system without transfer or termination reactions. They conclude that initially a Poisson distribution will form. This distribution will change with time until eventually the "most probable" distribution where M_w/M_n is two will be formed. For polystyrene polymerized anionically, the initial Poisson distribution is produced in a matter of seconds but the rate of depropagation is so slow that final equilibrium is not predicted for 100 years or so. As we have shown above, the depropagation of PTHF occurs very readily, and it can be polymerized under conditions that give a "reversible, living polymer system". There are suggestions in the literature that the molecular weight distributions predicted by BROWN and SZWARC (76) and by MIYAKE and STOCKMAYER (77) are being observed in the polymerization of THF.

In an initial study, MAKLETSOVA, EPEL'BAUM, ROZENBERG, and LYUDVIG (70) have looked at the molecular weight distributions of PTHF's resulting from bulk polymerizations initiated by trialkyloxonium salts and by antimony pentachloride at 20° C. They found that the two catalysts did indeed produce different molecular weight distributions (Fig. 20). The oxonium salt led to a $\overline{M}_w/\overline{M}_n$ of two, whereas the $\overline{M}_w/\overline{M}_n$ ratio observed after polymerization with SbCl$_5$, where initiation continues throughout polymerization, was 2.8. These data show that even with good initiation the system is tending toward a normal distribution at thermodynamic equilibrium.

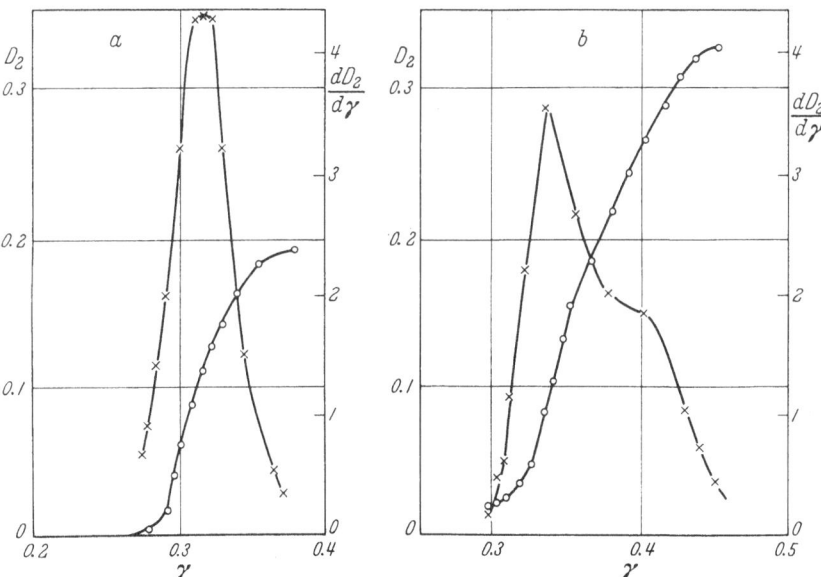

Fig. 20. Molecular weight distribution of PTHF: a-catalyst (C$_2$H$_5$)$_3$O$^+$BF$_4^-$; b-catalyst SbCl$_5$ (70)

In a more recent study, ROZENBERG, MAKLETSOVA, EPEL'BAUM et al. (74) have studied the variation of the molecular weight distribution at various degrees of conversion and in the presence of the chain transfer agent, water. They state that the method they used for determination of molecular weight distributions, turbidimetric titration, gives distribution figures that are of a qualitative, illustrative nature only. Thus they checked their results by calculating distributions from the $\overline{M}_w/\overline{M}_n$ ratios derived from light scattering and osmometry. The results from the two methods were in good agreement. Their results are shown in Fig. 21, which shows that as the degree of conversion increases, the distribution becomes broader and finally becomes symmetrical at thermodynamic equilibrium. The results are in agreement with the predictions (76, 77).

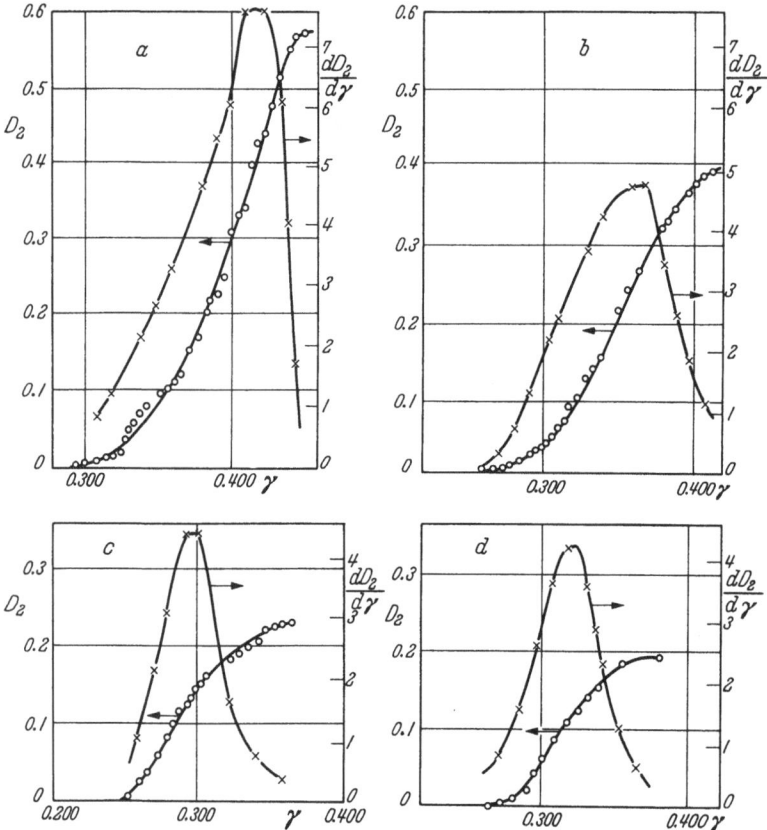

Fig. 21. Molecular weight distribution obtained by the turbidimetric method, as a function of the degree of conversion. Degree of conversion (mole/l): a —1.5; b —3.0; c —5.2; d —9.5 (74)

Additions of water during the polymerization of THF increased the polydispersity of the product polymer (Table 14). Even four percent of water (based on the catalyst) altered the molecular weight distribution of the equilibrium polymer due to the appearance of a considerable amount of low molecular weight fractions. At higher concentrations, the polydispersity was even greater.

VOFSI and TOBOLSKY (*32*) found good agreement at early stages of the polymerization between the degrees of polymerization estimated from C¹⁴

Table 14. *Influence of the addition of water on the molecular weight distribution of the equilibrium polymer* (*74*)

[H₂O] : [C]	M_w	M_n	M_w/M_n
0	300,000	156,000	2.00
0.04	190,000	57,000	3.34
	230,000	70,000	3.28 ★
0.33	140,000	20,000	7.00
	181,000	31,000	5.85 ★

★ Calculated from integral molecular weight distribution curves.

analysis $\overline{M}_n$ and $\overline{M}_w$, but noted some deviation at equilibrium conversion. Their results are shown in Fig. 22. They conclude that the molecular

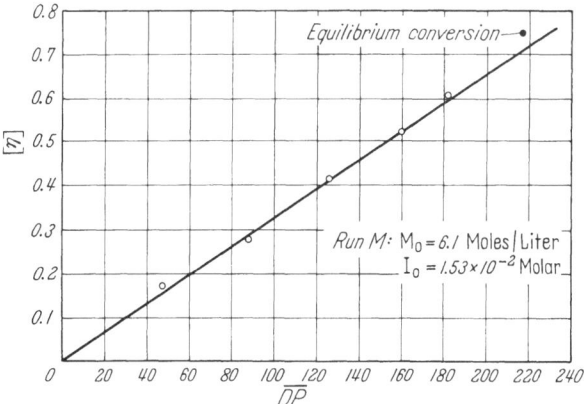

Fig. 22. Intrinsic viscosity vs. DP (from C¹⁴ analysis) (*32*)

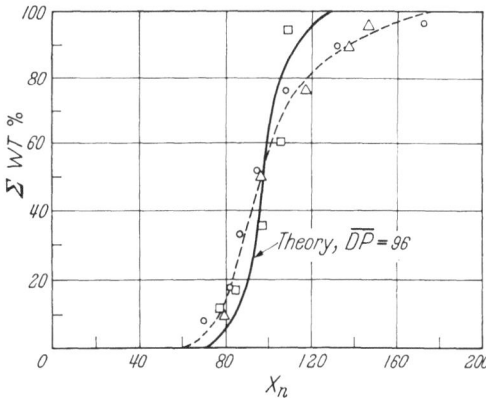

Fig. 23. Integral molecular weight (DP) distribution determined in triplicate for a PTHF of $\overline{DP}_n = 96$. Solid line is for a Poisson distribution of molecular weights (*34*)

weight distribution appears to be Poisson in the early stages and that it broadens in time as predicted (76, 77).

In a bulk polymerization study initiated with BF_3-ethylene oxide catalyst at $0°$ C, Ofstead (34) obtained quite narrow molecular weight distributions over the range from five to forty percent conversion. The comparison of his data for a PTHF of $\overline{DP}$ 96 with the theoretical curve for a Poisson distribution is shown in Fig. 23.

5. Unperturbed dimensions. Kurata, Utiyama, and Kamada (69) derived the unperturbed dimensions of PTHF in several solvents from the intrinsic viscosity data. Their results are given in Table 15.

Table 15. *End-to-end distance of PTHF in unperturbed state* (69)

σ^*	Solvent
1.85	benzene
1.70	cyclohexane
1.58	ethyl acetate
1.64	theta solvent **

  ★ σ is defined as the ratio of the unperturbed end-to-end distance to the end-to-end distance of a hypothetical chain with free internal rotation.
  ★★ Ethyl acetate (22.7 wt. %)/n-hexane (77.3 wt. %).

The effect of benzene on the chain conformation is notable.

May and Wetton (72) obtained another indication that benzene leads to an expansion of the polymer coils. They concluded from viscosity measurements that the unperturbed end-to-end distance in benzene was 1.34 times as great as that in the mixed theta solvent consisting of 39.2% heptane and 60.8% cyclohexane at $26°$ C.

6. Solubility parameter. Pass and Huglin (92) have measured the solubility parameter of PTHF in several ways. The results for a PTHF prepared with PF_5 are given in Table 16.

Table 16. *Solubility parameters of PTHF* (92)

Method	Solubility Parameter
Cross-linked PTHF swollen in aliphatic esters	8.6
[η] determinations in aliphatic esters	8.6
Swelling in conjunction with stress-strain measurements on cross-linked PTHF	8.5
By calculation from attraction constants of Small (Based on a density of 1.06 g/ml.)	8.85

B. Crystallization behavior

The study of the crystallization behavior of PTHF is just beginning. It is now well established from x-ray and other studies that PTHF is a crystalline polymer. It appears to have a planar zig-zag, but not fully linear chain conformation similar to polyethylene. This is in contrast to polyformaldehyde and polyethylene oxide which both have a helical conformation. The degree of crystallinity reported has varied from 33% (78) to 80% (79) depending both on the method of preparation of the polymer and the molecular weight. Lower molecular weights have higher degrees of crystallinity. The melting temperature usually reported is about 42° C (1, 78, 79, 83) although FARTHING (2) quotes a value of 58–60° C. SORENSON and CAMPBELL (80) give a value of 55° C for the high molecular weight polymer prepared with PF_5. TRICK (79) has estimated that the equilibrium melting temperature is about 51° C. There seems to be general agreement that the polymer glasses at about − 86° C (52, 78, 79, 81, 82).

1. The crystal structure of PTHF. By studying fiber photographs such as that shown in Fig. 24, CESARI, PEREGO, and MAZZEI (83) conclude that the unit-cell has a monoclinic symmetry, with $a = 5.61$ Å, $b = 8.92$ Å, $c = 12.25$ Å (fiber axis) and $\beta = 134°30'$; space group $C\,2/c$. They studied

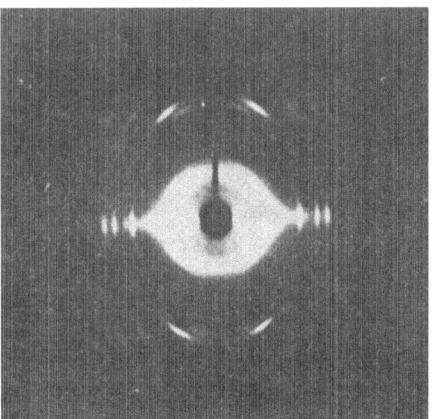

Fig. 24. Fiber photograph of PTHF (83)

a polymer prepared with Al alkyls-water/epichlorohydrin. IMADA, MIYAKAMA, CHATANI, TADOKORO and MURAHASHI (84) on the other hand obtain slightly different values on a PTHF prepared by using PF_5 as a catalyst: $a = 5.48$, $b = 8.73$, $c = 12.07$ Å (fiber axis), and $\beta = 134.2°$; space group $C\,2/c - C_{2h}$. They conclude that two molecular chains pass through the unit cell, that the molecular chain has a planar zig-zag conformation and that the zig-zag plane is parallel to the (100) plane.

The proposed arrangement of the polymer chains is shown in Figs. 25 and 26.

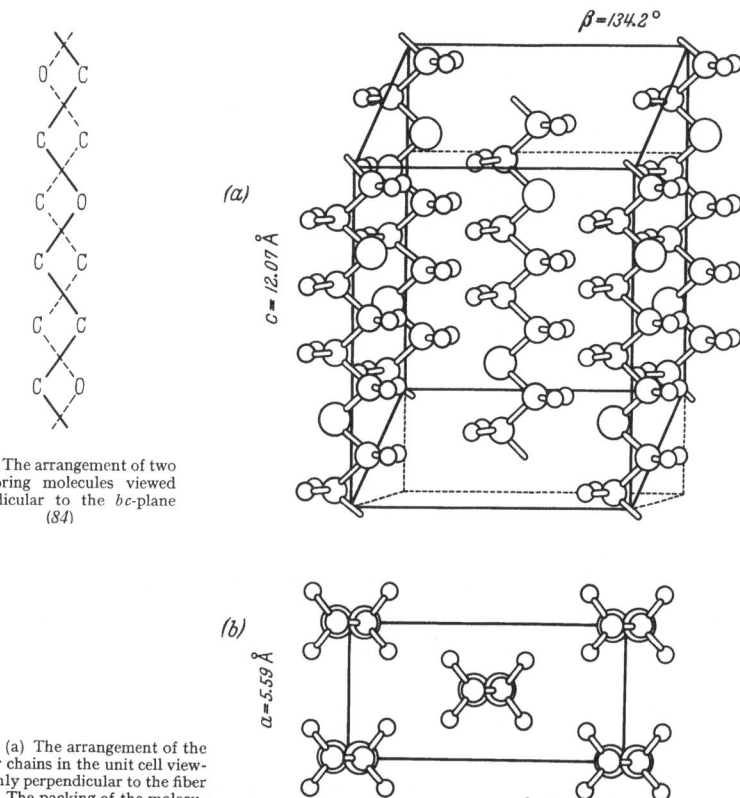

Fig. 25. The arrangement of two neighboring molecules viewed perpendicular to the bc-plane *(84)*

Fig. 26. (a) The arrangement of the polymer chains in the unit cell viewed roughly perpendicular to the fiber axis. (b) The packing of the molecules viewed down the fiber axis *(84)*

Further confirmation of the planar zig-zag structure has been obtained from a study of PTHF's infrared and Raman spectra *(84)* which are shown in Fig. 33.

2. Kinetics of crystallization. Trick *(79)* has reported a dilatometric study of the bulk crystallization of PTHF. The rates he observed for a polymer of $\overline{M_w} = 130,000$ (Polymer A) are shown in Fig. 27. He also found that a lower molecular weight polymer (Polymer B, $\overline{M_n} = 6760$) crystallized to a higher degree of crystallinity, whereas the introduction of comonomer units (Polymer C) decreased the degree of crystallinity (Fig. 28). From attempts to fit the Avrami Equation to the experimental data in the early stages of crystallization, a tentative value of $n = 3$ was assigned. This suggests the growth of instantane ously nucleated spherulites.

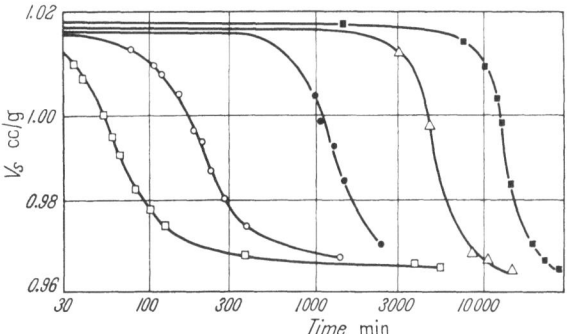

Fig. 27. Crystallization rates, Polymer A. □ 19.0° C, ○ 21.0° C, ● 24.0° C, △ 25.8° C, ■ 27.0° C (79)

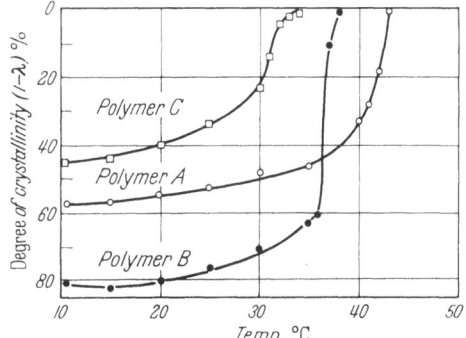

Fig. 28. Slow melting curves (79)

C. Degradation

1. Degradation in bulk. DAVIS and GOLDEN (85) have studied the degradation of PTHF in bulk at various temperatures. The polymers that they studied were prepared using a THF/PF$_5$ complex either in an open flask (polymer A) or in vacuum with exposure to air during the work up (polymer V). The intrinsic viscosity of polymer A, heated at fixed temperatures up to 150° C in a sealed system, fell rapidly to a constant value. Polymer V behaved similarly but the decrease was considerably smaller. When heated in air at a fixed temperature the viscosity of both polymers decreased continuously with eventual destruction of the polymer. Temperatures well in excess of 150° C were required for complete degradation of polymer A or V in vacuum.

The rate of weight loss of polymer A was determined in vacuum and in air. The activation energy of degradation in vacuum calculated from this was 45 kcal/mole. This value is comparable with those of polyethylene oxide (46 kcal/mole) but low compared to hydrocarbon polymers (60 to 70 kcal/mole). The corresponding activation energy of degradation in air was 29 kcal/mole.

KOVARSKAYA, LEVANTOVSKAYA, and YAZVIKOVA (86) recently studied the kinetics of thermooxidative degradation of PTHF of low molecular weight (1130) at 90—120° C. Hydroperoxides were formed at all stages of the oxidation. The rate of accumulation of peroxides reached a maximum of 2—2.3 millimoles/g at all temperatures. Kinetic curves of the oxidation revealed an auto-acceleration according to the formula:

$$- \Delta p = A \exp(\varphi t) , \qquad (49)$$

where Δp is the pressure change in the system, A is a constant, φ is the coefficient of autoacceleration, and t is time. The effective energy of activation of degradation that they found was 15 kcal/mole. The oxidation could be inhibited by the addition of amines and pyrocatechol.

2. Degradation in solution. DAVIS and GOLDEN (85) found that the viscosity of a benzene solution of PTHF heated in air dropped steadily in the absence of anti-oxidant with eventual complete destruction of the polymer. In the presence of anti-oxidant (e. g. 0.5% 2,6-di-t-butyl-4-methylphenol) the viscosity fell initially but thereafter became constant. As shown in Fig. 29, this initial drop was greater for polymer A

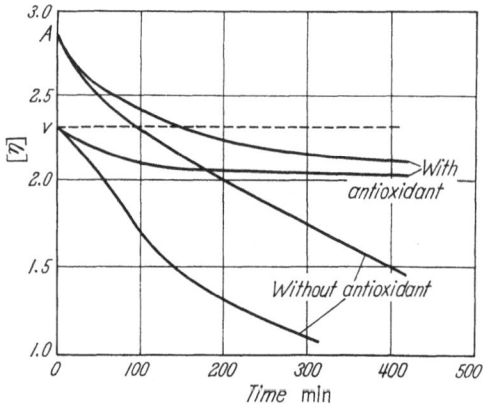

Fig. 29. Effect of time of heating on intrinsic viscosity, [η], of xylene solutions of PTHF prepared in air (A), prepared in vacuum (V). Polymer concentration 0.3%, anti-oxidant concentration 0.5% (85)

than for polymer V. In all cases degradation was accompanied by the appearance of a strong aldehydic carbonyl band in the infrared. DAVIS and GOLDEN (85) postulated that the weak links are peroxide groups formed during contact of the polymer with air. Indeed the polymers did contain peroxide groups that disappeared on heating. The mechanism proposed for degradation was

$$\sim CH_2\text{--}O\text{--}CH_2\sim \xrightarrow{O_2} \overset{\overset{\displaystyle OOH}{|}}{\underset{\downarrow}{CH\text{--}O\text{--}CH_2\sim}}$$
$$O\cdot + \cdot OH$$
$$\sim CH{=}O + \cdot OCH_2\sim \longleftarrow \sim \overset{|}{CH}\text{--}OCH_2\sim$$

(50)

At higher temperatures in the absence of air the degradation would be a random scission occurring primarily at the C–O bond of the PTHF:

$$RCH_2OCH_2R \rightarrow RCH_2O\cdot + \cdot CH_2R \qquad (51)$$

3. Degradation and cross-linking due to ionizing radiation. Analysis of the solubility behavior of PTHF after exposure to high energy electrons in vacuum (87) indicates a ratio of main chain scissions to cross-links of 0.37. These changes are accompanied by the evolution of H_2 and traces of other gases. The relationship between intrinsic viscosity and radiation dose is shown in Fig. 30. In this respect PTHF is similar to polyethylene

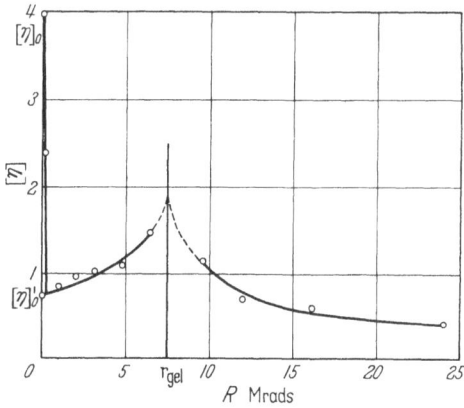

Fig. 30. Relationship between intrinsic viscosity $[\eta]$, and radiation dose (R, Mrads.) $[\eta]_0$: initial intrinsic viscosity, $[\eta]'_0$: effective initial intrinsic viscosity after destruction of weak links, r_{gel}: radiation dose at gel point. Ratio of chain scissions to cross-links is 0.37 (87)

oxide which also undergoes chain scission and crosslinking (Ratio 0.6). Since polymethylene oxide undergoes exclusive chain scission and polyethylene exclusively crosslinks, the scission probably takes place primarily at the weaker C–O bond. The ratio of scission to cross-linking is probably lower because the ratio of C–O to C–C is less in the case of PTHF than in polyethylene oxide.

D. Mechanical properties of PTHF

1. Mechanical properties compared to thermoplastics. BURROWS (88) compared a few of the mechanical properties of a BF_3-produced PTHF with typical values for some thermoplastics (Table 17).

Table 17. *Representative mechanical properties of various thermoplastics (88)*

Polymer	Tensile strength (psi × 10⁻³)	Modulus of elasticity (psi × 10⁻³)	Elongation %
PTHF	4.2	14	820
Polyethylene-low density	2.0	25	500
Polyethylene-linear	3.5	115	50
Polytetrafluoroethylene	4.0	580	200
Nylon 66	10.0	450	45
Poly BCMO	6.5	125	100

2. Mechanical properties compared to rubbers. A similar comparison of some high molecular weight PTHF's with some of the more familiar rubbers is shown in Table 18.

3. Dynamic mechanical measurements. Dynamic mechanical measurements (97) of the storage and loss components of the rigidity modulus (G', G'') at a single frequency are shown in Fig. 31. As is the case with all polyethers, there is a main relaxation region associated with the onset of microbrownian motion of the main chain. In the region of the melting temperature, a catastrophic drop in modulus appears.

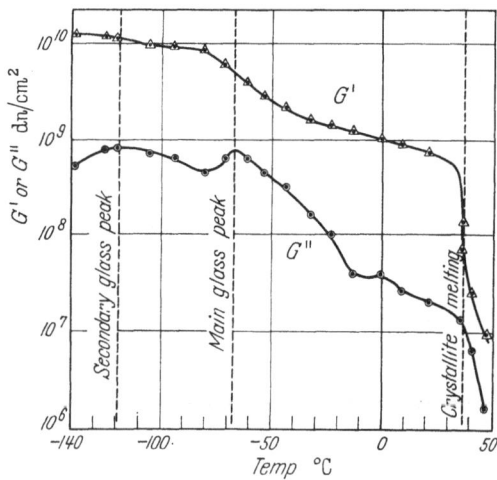

Fig. 31. Effect of temperature on the storage and loss components of the rigidity modulus of PTHF (97), $f = 100$ c/s

Fig. 32 shows the effect of varying the frequency. In this case the secondary mechanical γ process normally considered characteristic of a chain containing four or more methylenes is clearly visible.

Table 18. *Mechanical properties of some common rubber gum stocks compared to PTHF (89, 90, 91)*

Rubber	Shore A Hardness	Specific gravity (25° C)	Refractive index (20° C)	Dielectric constant (1 kc × sec, 25° C)	Tg, (°C)	Melting Temp. °C	Tensile strength (psi)	% Elongation
PTHF	95	0.99—1.18	1.48	5.0	—86	54	4000—6000*	300—600*
Hevea	30—100	0.92	1.52	2.7	—74	6(32)	>3000	780
SBR	40—100	0.94	1.53	2.9	—64	—35	<1000	270
Butadiene-Acrylonitrile	20—100	1.02	1.52	13.0	—23	—	>600	—
Neoprene	40—95	1.25	1.56	9.0	—50	32	>3000	875
Butyl	40—75	0.91	1.51	2.4	—75	—	>1500	675
Silicone	20—90	0.97	1.40	3.1	—123	—52	0	—

* Exact value depends on molecular weight.

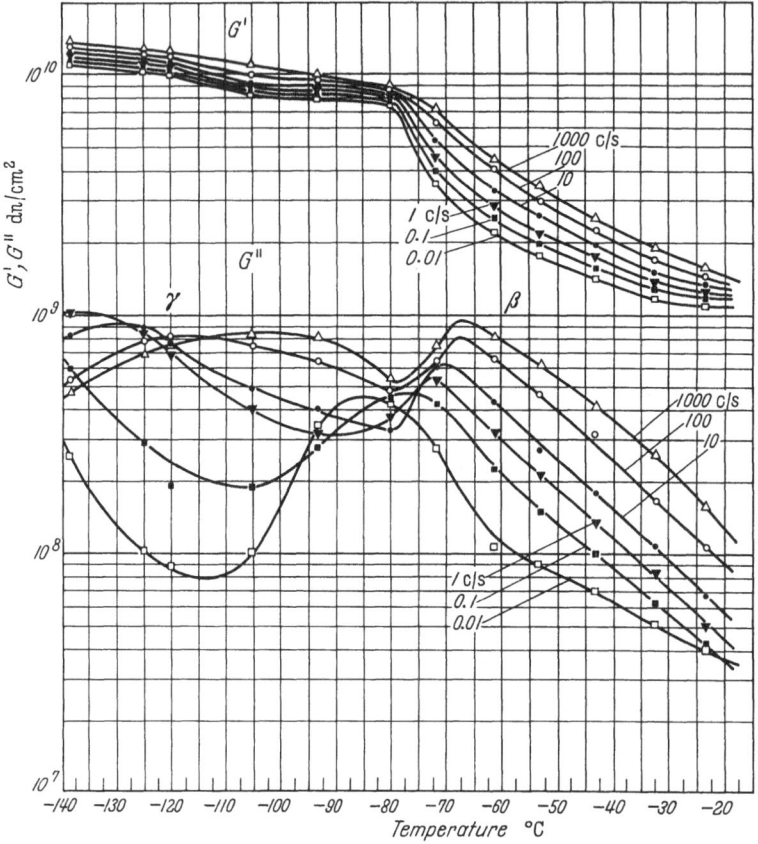

Fig. 32. Effect of frequency on the storage and loss components of the rigidity modulus of PTHF (97)

4. Mechanical properties of the vulcanizates. The vulcanization of PTHF using sulfur-peroxide systems has been reported (94, 95). Representative properties of some vulcanizates are given in Table 19.

Table 19. *Mechanical properties of cured PTHF Stock* (94)

Compound	Tensile strength (psi)	% Elongation	300% modulus (psi)
Unfilled	4000—5500	550—750	1200—1500
Black	4000—5000	600—770	1200—1800

E. Other properties

1. Permeability to salt. PTHF shows a certain resistance to the flow of salt (96) and has been suggested as a possible membrane for desalination experiments. Table 20 shows how PTHF compares with other polymers.

Table 20. *Resistance of polymers to the flow of salt (96)*

Polymer	Thickness μ	Water-flux λ/hr/cm²	Salt flux μg/hr/cm²	Salt rejection equivalent, %
Poly(vinyl isobutyl ether)	40	1.5	2.5	98.7
Poly(ethyl acrylate)	40	3.6	5.5	98.7
Polytetrahydrofuran	60	2.7	3.9	98.6
Poly(triallyl phosphate)	370	2.1	37.0	77
Poly(vinyl methyl ketone)	280	4.4	99.0	79
Cellulose triacetate	20	3.8	0.9	99.8

2. Spectra. Infrared, RAMAN, and NMR spectra of PTHF have been reported. IMADA et al. *(84)* have reported the Raman and polarized infrared spectra (Fig. 33). They made a tentative assignment of the bands.

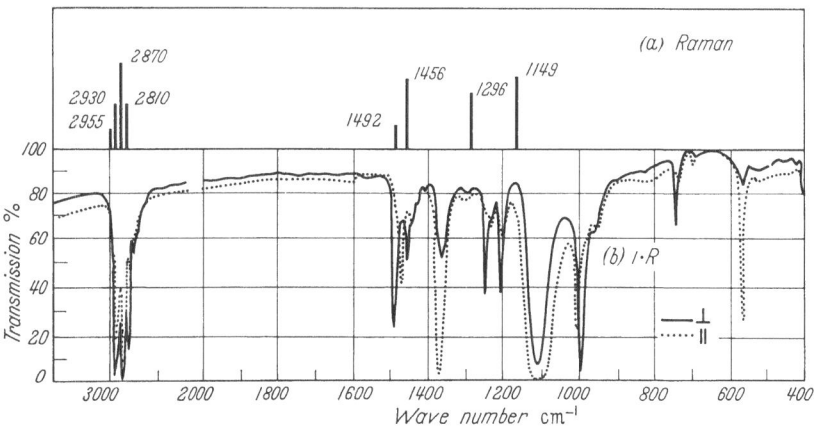

Fig. 33. (a) Raman spectrum and (b) polarized infrared spectra of PTHF. The solid (or broken) line in (b) represents the spectrum measured with the electric vector perpendicular (or parallel) to the orientation direction *(84)*

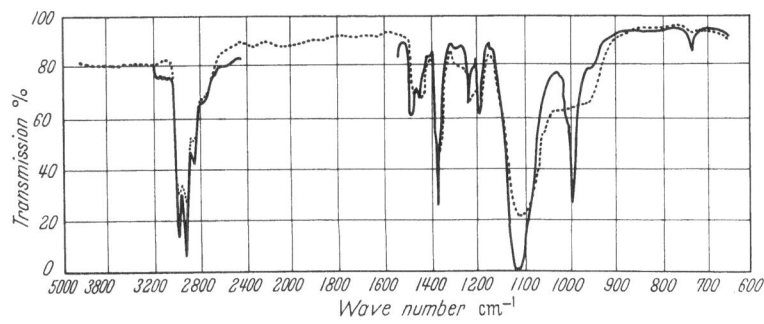

Fig. 34. Infrared spectrum of PTHF. (————): Crystalline (at 25° C); (- - - - -): Melt at (50° C) *(98)*

Saegusa, et al. (98) report the infrared spectrum of crystalline and amorphous PTHF (Fig. 34). The absorption band at 1000 cm⁻¹ was designated as one of the crystalline bands of PTHF. An NMR spectrum of PTHF in THF is shown by Kuntz (44).

3. Density. Huglin (93) has collected crystalline and amorphous densities from a variety of published and unpublished sources. These values are summarized in Table 21.

Table 21. *Density of PTHF (93)*

Type	T (°C)	Density (g/ml)
Amorphous	20	0.982
Amorphous	25	0.990
In solution	20	1.02
Crystalline (X-ray)	—	1.16—1.18
Crystalline (flotation)	—	1.07—1.08
Crystalline (weighing bottle)	—	1.06

VI. Copolymers and other polymers based on THF

A. Copolymerization with 1,2-epoxides and oxetanes

THF copolymerizes readily with other cyclic ethers such as oxides and oxetanes. The comonomers used include ethylene oxide (67), propylene oxide (99, 100), epichlorohydrin (ECH) (101, 102), phenyl glycidyl ether (102), 3.3-bis(chloromethyl) oxetane (BCMO) (25, 98, 101, 103) and 3-methyl-3-chloromethyl oxetane (103). Just as in THF homopolymerization, a large variety of catalysts have veen used. In many cases the kinetics of copolymerization have been studied. Table 22 summarizes the monomer reactivity ratios, r_1 (THF), and r_2 (comonomer) which have

Table 22. *Monomer reactivity ratios in THF copolymerization*

Comonomer	Catalyst	r_1 (THF)	r_2 (comonomer)	Ref.
Ethylene oxide	BF₃/ethylene oxide	2.2	0.08	67
ECH	AlEt₃–H₂O/ECH	20 ± 5	0.5 ± 0.3	101
ECH	AlEt₃/ECH	80 ± 20	2 ± 1	101
ECH	BF₃/ECH	3.9	0.06	36
ECH	BF₃/ECH	3.90	0.00	101
ECH	BF₃/ECH	3.85 ± 0.05	0.00 ± 0.05	108
ECH	BF₃/ECH	2.5	0.25	102
Phenylglycidyl ether	BF₃/ECH	$r_1 \approx r_2$		102
BCMO	BF₃	1.00 ± 0.05	0.82 ± 0.05	98
BCMO	AlEt₃–H₂O/ECH	1.50 ± 0.20	0.00	101
BCMO	AlEt₃/ECH	1.80 ± 0.01	0.01 ± 0.01	101
BCMO	AlEt₃–H₂O	0.40 ± 0.05	0.45 ± 0.05	103
3-Methyl-3-chloromethyl oxetane	AlEt₃	0.17 ± 0.1	1.74 ± 0.1	103

been reported. It is evident that there is some disagreement among different workers.

It is generally agreed that propagation in the cationic polymerization of cyclic ethers occurs after nucleophilic attack by the monomer oxygen atom (equation 3). Therefore, many authors attempt to explain their copolymerization data by noting that the more basic monomer has the higher reactivity with the active chain end. The order of basicity which has been established (36, 38) is:

$$\text{THF} > \text{BCMO} > 1,3\text{-dioxolane} > \text{propylene oxide} > \text{ECH} \,. \qquad (52)$$

This is broadly in agreement with the findings from copolymerization (Table 22 and Section VI-B). The introduction of comonomer units into PTHF reduces its crystallinity and rubbery materials generally result. In one case allylglycidyl ether was copolymerized so that the pendant olefinic side chains would allow sulfur vulcanization (1).

B. Copolymerization with cyclic formals

THF readily copolymerizes with cyclic formals also (25, 104). Okada et al. (104) have determined reactivity ratios for copolymerization of THF with 1,3-dioxolane using BF_3-etherate catalyst at $0°$ C. The values found are r_1 (THF) $= 28 \pm 4$, and r_2 (dioxolane) $= 0.25 \pm 0.05$.

Price and McAndrew (105) have copolymerized trioxane with THF using $Ph_3C^+SbCl_6^-$ as catalyst. The polymer was soluble in acetone and melted at $36-37°$ C.

C. Copolymerization with 7-oxabicyclo-[2:2:1]-heptane

A series of copolymers has been made from THF and 7-oxabicyclo-[2:2:1]-heptane using $FeCl_3$–$SOCl_2$ as catalyst. Low molecular weight polymers with melting points ranging from 150 to $320°$ C were reported (109).

D. Copolymerization with cyclic ethers that have no homopolymerizability

Certain cyclic ethers which will not homopolymerize will copolymerize with THF (25, 52). These cyclic ethers are stable five and six-membered ring compounds such as 2-methyltetrahydrofuran (2-MeTHF), and 1,4-dioxane (DOX). It is probable that 4-phenyl-1,3-dioxane (PhDOX), tetrahydropyran (THP), and 4-methyl-1,3-dioxolane (MDOL) which do not homopolymerize but which have been reported to copolymerize with BCMO (107, 108) would also copolymerize with THF. In the copolymerizations a correlation was again found between the basicity of the attacking ether and its reactivity with the cyclic oxonium ion. The order of basicities and the order of reactivity both were:

$$\text{2-MeTHF} > \text{THP} > \text{DOX} > \text{PhDOX} > \text{MDOL} . \qquad (53)$$

E. Copolymerization with other compounds

Copolymerizations of THF with styrene and with isobutylene have been reported in a patent (1). $CH_3COCl–AlCl_3$, $FeCl_3–POCl_3$, $SbCl_5$, and acetic anhydride-$HClO_4$ were employed as catalysts. Very viscous liquid polymers were reported.

Copolymerization of THF with diketene using $AlEt_3–H_2O$ catalyst at 0° C was recently described (106). The products were thought to be block copolymers instead of random copolymers.

F. Polyurethane elastomers from PTHF

PTHF is used in many urethane rubbers and as such it forms part of the backbone of some Spandex yarns. In these applications low molecular weight PTHF glycols are chain extended through their –OH end groups. The products take advantage of the rubbery properties and the high tensile strength of PTHF but destroy its tendency to crystallize by preventing long sequences of PTHF.

In a recent patent, REUTER (110) describes a polyurethane prepared from PTHF (mol. wt. 1000 to 3000), 1,4-butanediol, and $OCN(CH_2)_6CNO$. In another case MURBACH and ADICOFF (67) interrupted the regularity of PTHF by copolymerization with ethylene oxide before chain extension with diphenyl-methane-4,4'-disiocyanate. DICKINSON (99) prepared a series of polyurethane elastomers from THF-PO copolymer diols and 2,4-tolylene diisocyanate. He found that the use of copolymers with approximately 75 wt.-% THF led to polyurethanes with very good properties relative to the use of propylene oxide homopolymer.

VII. Conclusion

The study of PTHF and of the polymerization of THF is just beginning. Much interesting work, both theoretically and practically useful, remains to be done. Undoubtedly much of this is currently being done. It seems safe to suggest that THF will probably hold the place in the theoretical study of cationic polymerization that styrene now holds in the study of anionic polymerization.

Acknowledgement: We wish to thank Professor J. FURUKAWA, Dr. M. B. HUGLIN, Prof. T. SAEGUSA, Dr. D. SIMS, and Dr. R. E. WETTON for making available to us some of their papers prior to publication. We also wish to express our gratitude to Professor J. C. BAILAR JR, and Professor J. H. GOLDSTEIN for helpful discussions and valuable suggestions relating to portions of the manuscript.

VIII. References

1. FURUKAWA, J., and T. SAEGUSA: Polymerization of aldehydes and oxides. New York: Interscience Publishers 1963.
2. FARTHING, A. C.: Polymers from 1,3- and higher epoxides in high polymers. Vol. XIII, Chapter V. Polyethers Part I. Polyalkylene oxides and other polyethers. Ed. by N. G. GAYLORD. New York: Interscience Publishers 1963.
3. MEERWEIN, H., D. DELFS, and H. MORSHEL: Angew. Chem. 72, 927 (1960).
4. WURTZ, A.: Ann. Chim. Phys. 69, 330 (1863); as cited in Reference 1, p. 125.
5. MEERWEIN, H., D. DELFS, and H. MORSHEL: Angew. Chem 72, 927 (1960).
6. DAINTON, F. S., and K. J. IVIN: Quart Rev. 12, 61 (1958).
7. HALL JR., H. K.: ACS Polymer Preprints 6 (2), 535 (1965).
8. MEERWEIN, H.: German Patent 741,478 (June 21, 1939).
9. *U. S. Office of Technical Service Reports, Dept. of Commerce,* as enumerated in Reference 2, p. 310.
10. MEERWEIN, H., K. BODENBENNER, P. BORNER, F. KUNERT, and K. WUNDERLICH: Ann. 632, 38 (1960).
11. MILLER, D. B.: ACS Polymer Preprints 6 (2), 613 (1965).
12. TAKEGAMI, Y., T. UENO, and R. HIRAI: Bull. Chem. Soc. Japan 38, 1222 (1965); J. Polymer Sci. A-1 4, 973 (1966).
13. MUETTERTIES, E. L.: U. S. Patent 2,856,370 (Oct. 14, 1958).
14. MUETTERTIES, E., T. A. BUTLER, M. W. FARLOW, and D. D. COFFMAN: J. Inorg. Nucl. Chem. 16, 52 (1959).
15. SAEGUSA, T., H. IMAI, and J. FURUKAWA: Makromol. Chem. 65, 60 (1963).
16. WEISSERMEL, K., and E. NÖLKEN: Makromol. Chem. 68, 140 (1963).
17. BAWN, C. E. H., R. M. BELL, and A. LEDWITH: Chemical Society Anniversary Meeting, Cardiff, 1963.
18. — — — Polymer 6, 95 (1965).
19. — C. FITZSIMMONS, and A. LEDWITH: Proc. Chem. Soc. 1964, 391.
20. — R. M. BELL, C. FITZSIMMONS, and A. LEDWITH: Polymer 6, 661 (1965).
21. DREYFUSS, M. P., and P. DREYFUSS: Polymer 6, 93 (1965).
22. DAINTON, F. S., and K. IVIN: Quart. Rev. 12, 61 (1958).
23. SMALL, P. A.: Trans. Faraday Soc. 51, 1717 (1955).
24. ROZENBERG, B. A., O. M. CHEKHUTA, E. B. LYUDVIG, A. R. GANTMAKHER, and S. S. MEDVEDEV: Vysokomolekul. Soedin. 6 (11), 2030 (1964); Polymer Sci. USSR 6 (11), 2246 (1964).
25. DREYFUSS, M. P., and P. DREYFUSS: J. Polymer Sci. A-1 4, 2179 (1966).
26. SIMS, D.: J. Chem. Soc. 1964, 864.
27. FITZSIMMONS, C.: Private Communication.
28. IVIN, K. J., and J. LEONARD: Polymer 6, 621 (1965).
29. PLESCH, P. H.: The chemistry of cationic polymerization. Oxford: Pergamon 1963.
30. PRICE, C. C., and R. SPECTOR: J. Am. Chem. Soc. 88, 4171 (1966).
31. ROZENBERG, B. A., E. B. LYUDVIG, A. R. GANTMAKHER, and S. S. MEDVEDEV: Vysokomolekul. Soedin. 6 (11), 2035 (1964); Polymer Sci. USSR 6 (11), 2253 (1964).
32. VOFSI, D., and A. V. TOBOLSKY: J. Polymer Sci. A 3, 3261 (1965).
33. MEERWEIN, H., E. BATTENBERG, H. GOLD, E. PFEIL, and G. WILLFANG: J. Prakt. Chem. 154, 83 (1939).
34. OFSTEAD, E. A.: ACS Polymer Preprints 6 (2), 674 (1965).
35. SAEGUSA, T., H. IMAI, S. HIRAI, and J. FURUKAWA: Makromol. Chem. 54, 218 (1962).

36. Iwatsuki, S., N. Takigawa, M. Okada, Y. Yamashita, and Y. Ishii: J. Polymer Sci. B 2, 549 (1964).
37. Sims, D.: Makromol. Chem. 98, 245 (1966).
38. Wirth, H. E., and P. L. Slick: J. Phys. Chem. 66, 2277 (1962).
39. Ziegler, K.: In: Organometallic chemistry. Ed. by H. Zeiss. New York: Reinhold Publ. Corp. 1960.
40. Sims, D.: Makromol. Chem. 98, 235 (1966).
41. Lyudvig, E. B., B. A. Rozenberg, T. M. Zvereva, A. R. Gantmakher and S. S. Medvedev: Vysokomolekul. Soedin. 7 (2), 269 (1965); Polymer Sci. USSR 7 (2), 296 (1965).
42. Burrows, R. C., and B. F. Crowe: J. Appl. Polymer Sci. 6, 465 (162).
43. Kuntz, I.: ACS Polymer Preprints 7 (1), 187 (1966).
44. Kuntz, I.: J. Polymer Sci. B 4, 427 (1966).
45. Dreyfuss, M. P., J. C. Westfahl, and P. Dreyfuss: ACS Polymer Preprints 7 (2), 413 (1966).
46. Rose, J. B.: J. Chem. Soc. 1956, 542, 546.
47. Binks, J. H., and M. B. Huglin: Makromol. Chem. 93, 268 (1966).
48. Imai, H., T. Saegusa, S. Matsumoto, T. Tadasa, and J. Furukawa: Polymerization of tetrahydrofuran by triethyl-aluminum-water system. Makromol. Chem. (in press).
49. Berger, G., M. Levy, and D. Vofsi: J. Polymer Sci. B 4, 183 (1966).
50. Hermans, J. J.: J. Polymer Sci. 12 C, 345 (1966).
51. Rozenberg, B. A., E. B. Lyudvig, A. R. Gantmakher, and S. S. Medvedev: Vysokomolekul. Soedin. 7 (1), 188 (1965); Polymer Sci. USSR 7 (1), 205 (1965).
52. Dreyfuss, M. P., and P. Dreyfuss: Unpublished data.
53. Meerwein, H., G. Hinz, P. Hofmann, E. Kroning, and E. Pfeil: J. Prakt. Chem 147, 257 (1937).
54. Cotton, F. A., and G. Wilkinson: Advanced inorganic chemistry. p. 455. New York: Interscience Publishers 1963.
55. Pauling, L.: The nature of the chemical bond. 3rd Ed., p. 246. New York: Cornell University Press 1960.
56. Van Wazer, J. R.: Phosphorus and its compounds. Vol. 1, p. 241. New York: Interscience Publishers 1958.
57. Tophiev, A. V., S. V. Zavgorodnii, and Ya. M. Paushkin: BF_3 and its compounds as catalysts in organic chemistry. New York: Pergamon Press 1959.
58. Webster, M.: Chem. Rev. 66, 87 (1966).
59. Klages, F., H. Meuresch, and W. Steppich: Ann. 592, 81 (1955).
60. Lange, W. von, and K. Askitopoulos: Z. anorg. Chem. 223, 369 (1935).
61. Suschitzky, H.: J. Chem. Soc. 1953, 3042.
62. Schiemann, G., and W. Winkelmüller: In: Organic synthesis. Coll. Vol. 2, p. 188. Ed. by A. H. Blatt. New York: J. Wiley and Sons 1943.
63. Lange, W., and E. Müller: Ber. 63, 1058 (1930).
64. Balz, G., and G. Schiemann: Ber. 60, 1186 (1927).
65. Eastham, A. M.: Fortschr. Hochpolym. Forsch. 2, 18 (1960).
66. Saegusa, T., S. Matsumoto, T. Ueshima, and H. Imai: Preprint of paper: Polymerization of tetrahydrofuran by $AlEt_3 - H_2O$-promotor system, behavior of promotor; given at International Symposium on Macromolecular Chemistry, Japan, 1966.
67. Murbach, W. J., and A. Adicoff: Ind. Eng. Chem. 52, 772 (1960).
68. Ali, S. M., and M. B. Huglin: Makromol. Chem. 84, 117 (1965).
69. Kurata, M., H. Utiyama, and K. Kamada: Makromol. Chem. 88, 281 (1965).

70. MAKLETSOVA, N. V., I. V. EPEL'BAUM, B. A. ROZENBERG, and E. B. LYUDVIG: Vysokomolekul. Soedin. 7 (1), 70 (1965); Polymer Sci. USSR 7 (1), 73 (1965).
71. EVANS, J. M., and M. B. HUGLIN: Private Communication.
72. MAY, R. W., and R. E. WETTON: Private Communication.
73. FLORY, P. J.: Principles of polymer chemistry, p. 282. Ithaca, New York: Cornell University Press 1953.
74. ROZENBERG, B. A., N. V. MAKLETSOVA, I. V. EPEL'BAUM, E. B. LYUDVIG, and S. S. MEDVEDEV: Vysokomolekul. Soedin. 7 (6), 1051 (1965); Polymer Sci. USSR 7 (6), 1163 (1965).
75. HACHIHAMA, Y., and T. SHONO: Technol. Reports Osaka Univ. 9, No. 361, 299 (1958).
76. BROWN, W. B., and M. SZWARC: Trans. Faraday Soc. 54, 416 (1958).
77. MIYAKE, A., and W. H. STOCKMAYER: Makromol. Chem. 88, 90 (1965).
78. WETTON, R. E., and G. WILLIAMS: Trans. Faraday Soc. 61, 2132 (1965).
79. TRICK, G. S., and J. M. RYAN: ACS Polymer Preprints 7 (1), 92 (1966).
80. SORENSON, W. R., and T. W. CAMPBELL: Preparative methods of polymer chemistry, p. 255. New York: Interscience Publications 1961.
81. WILLBOURN, A. H.: Trans. Faraday Soc. 54, 717 (1958).
82. BRANDRUP, J., and E. H. IMMERGUT: Polymer Handbook, p. III-80. New York: Interscience Publishers 1966.
83. CESARI, M., G. PEREGO, and A. MAZZEI: Makromol. Chem. 83, 196 (1965).
84. IMADA, K., T. MIYAKAWA, Y. CHATANI, H. TADAKORO, and S. MURAHASHI: Makromol. Chem. 83, 113 (1965).
85. DAVIS, A., and J. H. GOLDEN: Makromol. Chem. 81, 38 (1965).
86. NEIMAN, M. B., B. M. KAVARSKAYA, I. I. LEVANTOVSKAYA, and M. P. YAZVI-KOVA: Plasticheskie Massy. 1966 (1), 42, through C. A. 64, 12827 h (1966).
87. GOLDEN, J. H.: Makromol. Chem. 81, 51 (1965).
88. BURROWS, R. C.: ACS Polymer Preprints 6 (2), 600 (1965).
89. WINSPEAR, G. G.: Vanderbilt Rubber Handbook. New York: R. T. Vanderbilt Co. Inc. 1958.
90. Compounders Technical Guide, CTG No. 49, 1. The B. F. Goodrich Co., Research Center, Brecksville, Ohio.
91. HODGMAN, C. D.: Handbook of Chemistry and Physics. Cleveland, Ohio: The Chemical Rubber Publishing Co. 1960.
92. PASS, D., and M. B. HUGLIN: Private Communication.
93. HUGLIN, M. B.: Private Communication.
94. Goodyear Tire and Rubber Co., British patent 1,006,316 (Cl. C 08g), Sept. 29, 1965.
95. Goodyear Tire and Rubber Co., British patent 1,001,345 (Cl. C 08g), Aug. 18. 1965.
96. BLOCK, R., O. KEDEM, and D. VOFSI: J. Polymer Sci. B 3, 965 (1965).
97. WETTON, R. E.: British Polymer Physics Group Meeting, Shrivenham, 1965.
98. SAEGUSA, T., H. IMAI, and J. FURUKAWA: Makromol. Chem. 56, 55 (1962).
99. DICKINSON, L. A.: J. Polymer Sci. 58, 857 (1962).
100. KURENGINA, T. N., L. V. ALVEROVA, and V. A. KROPATCHEY: Vysokomolekul. Soedin. 8 (2), 293 (1966).
101. SAEGUSA, T., T. UESHIMA, H. IMAI, and J. FURUKAWA: Makromol. Chem. 79, 221 (1964).
102. ROZENBERG, B. A., E. B. LYUDVIG, N. V. DESYATOVA, A. R. GANTMAKHER, and S. S. MEDVEDEV: Vysokomolekul. Soedin. 7 (6), 1010 (1965); Polymer Sci. USSR 7 (6), 1116 (1965).

103. ALFEROVA, L. V., and V. A. KROPACHEV: Vysokomolekul. Soedin. 7 (6), 1065 (1965); Polymer Sci. USSR 7 (6), 1177 (1965).
104. OKADA, M., N. TAKIKAWA, S. IWATSUKI, Y. YAMASHITA, and Y. ISHII: Makromol. Chem. 82, 16 (1965).
105. PRICE, M. B., and F. B. McANDREW: ACS Polymer Preprints 7 (1), 207 (1966).
106. TADA, K., Y. YAMADA, T. SAEGUSA, and J. FURUKAWA: Polymer Previews 2, 174 (1966) (Makromol. Chem., in press).
107. TSUDA, T., T. NOMURA, and Y. YAMASHITA: Makromol. Chem. 86, 301 (1965).
108. ISHIGAKI, A., T. SHONO, and Y. HACHIKAMA: Makromol. Chem. 79, 170 (1964).
109. WITTBECKER, E. L., H. K. HALL, and T. W. CAMPBELL: J. Am. Chem. Soc. 82, 1218 (1960).
110. REUTER, G.: Neth. Appl. 6,404,809 (Cl. C 08g), July 26, 1965; Ger. Appl. Jan. 25, 1964; through C. A. 64, 5276d (1966).

Received September 19, 1966

SPRINGER-VERLAG
BERLIN HEIDELBERG GMBH

Chemie, Physik und Technologie
der Kunststoffe in Einzeldarstellungen

Herausgegeben von K. A. Wolf

9. Band

Rauch-Puntigam/Völker:
Acryl- und Methacryl-
verbindungen

Von Dr. phil. H. Rauch-Puntigam, Forschungslaboratorium
der Vianova-Kunstharz AG, Graz/Österreich
und Dr. phil. Th. Völker, Forschungslaboratorium
der Lonza AG in Fribourg/Schweiz
Mit 24 Abbildungen. XX, 429 Seiten Gr.-8°. 1966
Ganzleinen DM 89,—

10. Band

Voigt:
Die Stabilisierung der Kunst-
stoffe gegen Licht und Wärme

Von Dr. rer. nat. Joachim Voigt, Farbwerke Hoechst AG.,
vormals Meister Lucius & Brüning, Frankfurt/Main-Hoechst
Mit 36 Abbildungen. XII, 643 Seiten Gr.-8°. 1966
Ganzleinen DM 98,—

11. Band

Wandel/Tengler/Ostromow:
Die Analyse von Weichmachern

Von Dr. rer. nat. Martin Wandel, Dr. rer. nat.
Hubert Tengler, beide Farbenfabriken Bayer AG, Dormagen
und Dipl.-Chem. Hermann Ostromow,
Farbenfabriken Bayer AG, Leverkusen
Mit 126 Abbildungen. Etwa 200 Seiten Gr.-8°. 1966
Ganzleinen etwa DM 38,—